Lecture Notes in Statistics 107

Edited by P. Bickel, P. Diggle, S. Fienberg, K. Krickeberg, I. Olkin, N. Wermuth, S. Zeger

W0042836

Springer
New York
Berlin
Heidelberg
Barcelona
Budapest
Hong Kong
London
Milan
Paris
Tokyo

Masafumi Akahira
Kei Takeuchi

Non-Regular Statistical Estimation

 Springer

Masafumi Akahira
Institute of Mathematics
University of Tsukuba
Ibaraki 305
Japan

Kei Takeuchi
Faculty of International Studies
Meiji-Gakuin University
Kamikuramachi 1598
Totsukaku, Yokohama 244
Japan

Library of Congress Cataloging-in-Publication Data Available
Printed on acid-free paper.

9 8 7 6 5 4 3 2 1

ISBN-13: 978-0-387-94578-1   e-ISBN-13: 978-1-4612-2554-6

DOI: 10.1007/978-1-4612-2554-6

Preface

In the theory of statistical estimation, there are well known results such as Cramér-Rao inequality for the variance of unbiased estimators, the asymptotic normality and efficiency of the maximum likelihood estimators etc. in "usual" situations. What the adjective "usual" means is stated in a set of regularity conditions which are more or less rigorously dealt with in higher than elementary level text books, but are often considered to be troublesome details only for the sake of mathematical exactness. Then, there are less well known "surprising" results such as the existence of zero variance unbiased estimators, "superefficient" estimators etc. in one or other "unusual" situations. These are sometimes mentioned in more advanced textbooks and monographs, but are dealt as rather "pathological" cases.

The fact that "unusual" or "non-regular" cases are not necessarily "pathological" seems to have been noted by very few authors. From one viewpoint "regular" cases might be considered to be the "boundary" situation among all possible cases, therefore rather "special" cases, if not "exceptional" cases. For example, for independent and identically distributed cases, the maximum order of consistency of estimators can be $n^{1/\alpha}$ ($0 < \alpha \leq 2$) and $\sqrt{n \log n}$, and is $\sqrt{n}$ for the regular case, and never be of $n^{1/\alpha}$ for $\alpha > 2$. Also, it is a rule rather than an exception that we have no asymptotically efficient estimators in general situations.

Thus the set of regularity conditions which are considered to represent "usual" situation are not mathematical trivialities but are really restricting conditions to get the "usual" results. Hence it would be valuable to check each set of regularity conditions to clarify what happens if it does not hold.

The present monograph is to show the meaning and implications of regularity conditions as systematically as possible. There seems to have been few books and monographs which treated non-regular cases systematically, for asymptotic theory, Ibragimov and Has'minskii (1981) is an outstanding example. This monograph is based on a series of our papers which have been published sporadically over a long period and in scattered places. Our previous monograph of Vol.7 of this series on asymptotic estimation, also dealt with "general" cases including a non-regular situation, but not necessarily focussed on "non-regular cases." Still many results in non-regular asymptotic cases were discussed in the previous monograph, hence in this one emphasis is laid on small sample results as well as asymptotics.

Contents

CHAPTER 1

GENERAL DISCUSSIONS ON UNBIASED ESTIMATION

1.1. FORMULATION

"Non-regular estimation" literally means the theory of statistical estimation when some or other of the regularity conditions for the "usual" theory fail to hold. The concept itself is purely negative and it seems to be almost self-contradiction to try to establish a "general theory" of non-regular estimation. In small sample and large sample theories of estimation of real parameters, however, there are well established sets of regularity conditions, and it is worth while to examine what may follow if any one of these regularity conditions fails to hold. And there has been accumulated substantial amount of results obtained by many authors, though somewhat sporadic investigations, which can give some insight into the structure of non-regular estimation and can clarify the "meaning" of each of the regularity conditions by showing which part of the theorem fails to hold and how it must be modified if it is not satisfied. The purpose of this chapter is to review those results from some unifying viewpoint and also to point out some problems yet to be solved. Our main interest is, therefore, not to look for some strange looking "pathological" examples, but rather to contribute to the main stream of the theory of statistical estimation by clarifying the "regular" theory from the reverse side. Many of the results given in this chapter are discussed in more detail in the subsequent chapters.

First we shall consider the set of regularity conditions of statistical estimation theory. In most general abstract terms, the problem of statistical estimation can be formulated as follows. Let $(\mathcal{X}, \mathcal{A})$ be a sample space and let $\{P_\omega \,|\, \omega \in \Omega\}$ be a set of probability measures over $(\mathcal{X}, \mathcal{A})$. The index set Ω can be any abstract set. Let $\theta = \theta(\omega)$ be a real p-dimensional vector valued function of ω and is called the parameter. An estimator $\hat{\theta} = \hat{\theta}(x)$ is a measurable function from $\mathcal{X}$ into Euclidean p-space $\mathbf{R}^p$. An estimator which always comes close to $\theta(\omega)$ when a random variable X is distributed according to P_ω, $\omega \in \Omega$, is considered to be a "good" estimator. The main set of regularity conditions usually considered as follows :

(A1) The set of probability measures is dominated by a σ-finite measure μ over $(\mathcal{X}, \mathcal{A})$, and the density function $dP_\omega/d\mu = f(x, \omega)$ $(\omega \in \Omega)$ is defined.

(A2) For each $\omega \in \Omega$, the support of P_ω, i.e. the set $\{x \,|\, f(x, \omega) > 0\}$ can be defined independently of ω.

(A3) The index space Ω itself is an open subset of $\mathbf{R}^q$, where $q \geq p$.

(A4) For almost all $x[\mu]$, $f(x, \omega)$ is continuous in ω.

(A5) For almost all $x[\mu]$, $f(x, \omega)$ is (k-times) continuously differentiable with respect to ω.

(A6) The Fisher information matrix

$$I_\omega = E_\omega \left[\frac{\partial}{\partial \omega} f(X, \omega) \frac{\partial}{\partial \omega'} f(X, \omega) \right], \quad \omega \in \Omega$$

is well defined and finite.

For large sample theory we have to consider a sequence of sample spaces $(\mathcal{X}_{(n)}, \mathcal{A}_{(n)})$ ($n = 1, 2, \ldots$) and that of probability measures $P_\omega^{(n)}$ ($\omega \in \Omega$) on $(\mathcal{X}_{(n)}, \mathcal{A}_{(n)})$ ($n = 1, 2, \ldots$) with the common index space Ω, and additional regularity conditions are the following.

(L1) For each n, $(\mathcal{X}_{(n)}, \mathcal{A}_{(n)})$ is subset of $(\mathcal{X}_{(n+1)}, \mathcal{A}_{(n+1)})$ and $P_\omega^{(n)}$ ($\omega \in \Omega$) is the marginal probability measure over $(\mathcal{X}_{(n)}, \mathcal{A}_{(n)})$ derived from the probability measure $P_\omega^{(n+1)}$ ($\omega \in \Omega$) over $(\mathcal{X}_{(n+1)}, \mathcal{A}_{(n+1)})$. (Sequence of observations).

(L2) For each n, $(\mathcal{X}_{(n)}, \mathcal{A}_{(n)})$ is the n-fold direct product space of sample spaces $(\mathcal{X}_i, \mathcal{A}_i)$ ($i = 1, \ldots, n$) and $P_\omega^{(n)}$ ($\omega \in \Omega$) is the corresponding product measure $P_\omega^{(n)}$ ($\omega \in \Omega$) on $(\mathcal{X}_{(n)}, \mathcal{A}_{(n)})$ of P_ω^i ($\omega \in \Omega$) on $(\mathcal{X}_i, \mathcal{A}_i)$ ($i = 1, \ldots, n$). (Independence).

(L3) For each $i = 1, \ldots, n$, $\mathcal{X}_i = \mathcal{X}, \mathcal{A}_i = \mathcal{A}$ and $P_\omega^i = P_\omega^1$ ($\omega \in \Omega$). (Identical distribution).

Another set of regularity conditions are :

(S1) The probability measures P_ω ($\omega \in \Omega$) admit a finite dimensional sufficient statistic $T = t(X)$.

(S2) The sufficient statistic is complete.

(S3) The probability measures form an exponential family, i.e., $f(x, \omega)$ can be expressed as

$$f(x, \omega) = c(\omega) h(x) \exp\{s(\omega)' t(x)\},$$

where $s(\omega)$ and $t(x)$ are p-dimensional real vectors.

There are still some other minor conditions for various theorems, but the most commonly discussed are included in the above.

Next we consider the theorems of estimation. For small sample situation, most of the theory deals with unbiased estimation, and main theorems deal with (a) the existence of locally best unbiased estimators and (b) the existence of uniformly minimum variance unbiased (UMVU) estimators. Also in small sample situation we may consider other types of "unbiasedness" condition than the usual expectation-unbiasedness, and other types of dispersion criteria than the variance. For large sample theory, the main results are concerned with (a) the existence of consistent estimators, (b) the maximum order of consistency, (c) the asymptotic efficiency of estimators, and (d) higher order asymptotic efficiency of estimators (see, e.g., Akahira and Takeuchi, 1981; Ibragimov and Has'minskii, 1981; Lehmann, 1983; Pfanzagl and Wefelmeyer, 1985; Amari, 1985; Ghosh, 1994; Barndorff-Nielsen and Cox, 1994).

In the subsequent sections we consider various combinations of problems and situations arising from failures of the above conditions.

1.2. UNDOMINATED CASE

The most extremely non-regular case would be the one when the condition (A1) is not satisfied, i.e., the probability measures are not dominated. There exist, however, rather simple examples of the undominated case.

Example 1.2.1. Let $\Omega = \{\omega_1, \ldots, \omega_N\}$ be a collection of N real values. The sample consists of the pair J and Y, where J is a set of integers $(I_1, \ldots, I_n)$ with $1 \leq I_i \leq N$ $(i = 1, \ldots, n)$ and Y is a set of n real values $(Y_1, \ldots, Y_n)$. And the probability measure is defined as

$$P\{(I_1, \ldots, I_n) = (i_1, \ldots, i_n)\} = p(i_1, \ldots, i_n)$$

and $Y_i = \omega_{I_i}$ $(i = 1, \ldots, n)$ with probability 1 and it is assumed that $p(i_1, \ldots, i_n)$ is independent of ω. This is nearly most general formulation of the problem of survey sampling when Ω denotes the set of the values of a "finite population" and Y the set of sample values, and P the "sampling scheme". The problem is to estimate some "population" parameter $\theta = \theta(\omega_1, \ldots, \omega_N)$. In general case the sample size n may be random, and I_i's need not all be distinct. (The case of sampling with replacement.) If we assume that ω_i can be any element of an open subset of $\mathbf{R}^1$, the distribution is not dominated. Then the following is known (Takeuchi (1961a), Cassel et al. (1977)).

Theorem 1.2.1. *For parameter $\theta = \theta(\omega)$ with unbiased estimators, a locally minimum variance unbiased estimator at specified $\omega = \omega_0$ has zero variance at $\omega = \omega_0$. Hence there never exists an UMVU estimator unless there is one which is always of zero variance.*

Proof. Suppose that $\hat{\theta}$ is an unbiased estimator of $\theta(\omega)$. Then $\hat{\theta}$ is expressed as

$$\hat{\theta} = \hat{\theta}(I, Y) = \hat{\theta}(I_1, \ldots, I_n, Y_1, \ldots, Y_n).$$

And from the unbiasedness condition we have

$$E_\omega(\hat{\theta}) = \sum_{(i_1, \ldots, i_n)} p(i_1, \ldots, i_n)\hat{\theta}(i_1, \ldots, i_n, \omega_{i_1}, \ldots, \omega_{i_n}) = \theta(\omega),$$

where the sum is all over n-tuples $(i_1, \ldots, i_n)$ from the integers $1, \ldots, N$. Now let $\omega_0 = (\omega_1^0, \ldots, \omega_N^0)$ be arbitrarily fixed, and put

$$\hat{\theta}_0 := \hat{\theta}(I_1, \ldots, I_n, Y_1, \ldots, Y_n) - \hat{\theta}(I_1, \ldots, I_n, \omega_{I_1}^0, \ldots, \omega_{I_n}^0) + \theta(\omega_0).$$

Then it immediately follows that if $\omega = \omega_0$, $\hat{\theta} \equiv \theta(\omega_0)$. Hence $V_{\omega_0}(\hat{\theta}_0) = 0$, and for any ω

$$E_\omega(\hat{\theta}_0) = E_\omega(\hat{\theta}) - \sum_{(i_1, \ldots, i_n)} p(i_1, \ldots, i_n)\hat{\theta}(i_1, \ldots, i_n, \omega_{i_1}^0, \ldots, \omega_{i_n}^0) + \theta(\omega_0)$$

$$= \theta(\omega) - \theta(\omega_0) + \theta(\omega_0)$$

$$= \theta(\omega).$$

Therefore $\hat{\theta}_0$ is unbiased and has zero variance when $\omega = \omega_0$.

Example 1.2.2. Let $X_1, \ldots, X_n$ be independently and identically distributed (i.i.d.) and for each $i = 1, \ldots, n$, $X_i - \alpha$ $(0 \le \alpha < 1)$ is distributed according to the Poisson distribution with parameter λ, where n is fixed. The index is the pair of non-negative constants α and λ. The class of distributions is clearly undominated, but $([\bar{X}], \bar{X} - [\bar{X}])$ is obviously sufficient and complete, where $[s]$ denotes the largest integer less than or equal to s, and UMVU estimator exists for any estimable parameter.

Example 1.2.3. Let $X_1, \ldots, X_n$ be i.i.d. random variables. The class of possible distributions P of X_1 is the set of the all discrete distributions over the real line. Obviously the class is not dominated. Now let $X_{(1)} \le \cdots \le X_{(n)}$ be the order statistic obtained from $X_1, \ldots, X_n$. We denote $(X_{(1)}, \ldots, X_{(n)})$ by Y. Y is sufficient, and we shall prove that it is complete. Assume that

$$E\left[\phi(Y)\right] = 0 \quad \text{for all discrete distributions } P.$$

Now take, as any possible value of Y, $y = (y_1, \ldots, y_n)$, $y_1 \le \cdots \le y_n$ of which there are k distinct values $z_1 < \cdots < z_k$ and the numbers of y_i values equal to z_j be n_j, $(\sum_{j=1}^{k} n_j = n)$. Now consider the class M_z of discrete distributions with the support $(z_1, \ldots, z_k)$ and $P(X = z_j) = p_j$ $(j = 1, \ldots, k)$, $p_j > 0$ $(j = 1, \ldots, k)$, $\sum_{j=1}^{k} p_j = 1$. Then from completeness of multinomial distributions we have $E\left[\phi(Y)\right] = 0$ for all $p \in M_z$ implies $\phi(y) \equiv 0$ over the set of values of Y where all y_i's are equal to one of z_j's, and this implies that $\phi(y) = 0$. Since y can be taken arbitrarily, we have $\phi(y) = 0$ for all y, which completes the proof.

Consequently from the case any parameter with unbiased estimators admits a UMVU estimator.

The above examples show the undominated cases where some uniform results for the existence of UMVU estimators can be established. There is also a simple but rather strange example of the undominated case as shown below.

Example 1.2.4. Suppose that for every $n \ge 2$, $X_1, \ldots, X_n$ are i.i.d. random variables according to the probability distribution with the probability mass equal to $1/2$ concentrated at the point θ, and the rest uniformly distributed over the interval $(0, 1)$. The parameter θ is assumed to be unknown in the interval $(0, 1)$. The class of probability distributions is undominated, but we can construct an unbiased estimator by the following : Let θ_0 be any constant in the interval $(0, 1)$. When two or more of X_i's have identical values, let the value be X^* and

$$\hat{\theta}_n = \frac{1}{1 - (n+1)2^{-n}}(X^* - \theta_0) + \theta_0,$$

and $\hat{\theta}_n = \theta_0$ otherwise. Then it can be shown that $E_\theta(\hat{\theta}_n) = \theta$ for all θ and it is also obvious that $V_{\theta_0}(\hat{\theta}_n) = 0$. Hence the variance of the locally minimum variance unbiased (LMVU) estimator is equal to zero.

This example may be considered to be the limiting case of the regular situation with the following density $f(x, \theta)$ as ε tends to zero :

$$f(x, \theta) = \begin{cases} \dfrac{1}{2} & \text{for} \quad |x - \theta| > \varepsilon, \ 0 < x < 1, \\ \dfrac{1}{2} + \dfrac{1}{2\varepsilon}\left(1 - \dfrac{|x - \theta|}{\varepsilon}\right) & \text{for} \quad |x - \theta| \leq \varepsilon, \ 0 < x < 1 \\ 0 & \text{otherwise}, \end{cases}$$

where $0 < \varepsilon < \theta < 1 - \varepsilon$. Then the amount I_ε of the information is given by

$$\begin{aligned} I_\varepsilon &= \int_0^1 \left\{\frac{\partial}{\partial \theta} \log f(x, \theta)\right\}^2 f(x, \theta) dx \\ &= \int_{\theta - \varepsilon}^{\theta + \varepsilon} \frac{1}{4\varepsilon^4 f(x, \theta)} dx = \frac{1}{2\varepsilon^4} \int_{-1}^1 \frac{\varepsilon}{1 + \frac{1}{\varepsilon}(1 - |t|)} dt \\ &= \frac{1}{\varepsilon^2} \int_0^1 \frac{1}{\varepsilon + 1 - t} dt = \frac{1}{\varepsilon^2} \log \frac{1 + \varepsilon}{\varepsilon}. \end{aligned}$$

Hence I_ε tends to infinity as $\varepsilon \to 0$, which implies that the lower bound of the variance of unbiased estimators goes to zero.

1.3. THE SUPPORT DEPENDING ON THE PARAMETER

In the case when the support of the distribution depends on the parameter, we often encounter a situation where the variance of locally minimum variance unbiased (LMVU) estimator is zero (see Section 2.2). In other cases we have the infimum of the variance of unbiased estimator is zero, but there does not exist one with zero variance (see Sections 2.2 to 2.4).

Example 1.3.1. Let X be a random variable with a density function $f(x, \theta)$, where θ is a real-valued parameter. We consider the location parameter case, i.e. $f(x, \theta) = f_0(x - \theta)$. Assume that

$$\begin{aligned} f_0(x) > 0 & \quad \text{for} \quad a < x < b \ ; \\ f_0(x) = 0 & \quad \text{for} \quad x \leq a, \ x \geq b, \end{aligned}$$

and further that $f_0(x)$ is continuously differentiable in the open interval (a, b) and

$$\lim_{x \to a+0} f_0(x) = A_0, \quad \lim_{x \to b-0} f_0(x) = B_0$$

exist, including the case where either or both A_0 and B_0 are infinity. Then it is shown that

$$\min_{\hat{\theta}\,:\,\text{unbiased}} V_{\theta_0}\left(\hat{\theta}(X)\right) = 0$$

for any specified value of θ_0. Here an estimator $\hat{\theta}_0$ with zero variance at $\theta = \theta_0$ can be constructed as follows. Let $\hat{\theta}_0(x) = \theta_0$ for $\theta_0 + a < x < \theta_0 + b$. And for $\theta_0 + b < x \leq \theta_0 + 2b - a$, $\hat{\theta}_0(x)$ is determined by the equation

$$(1.3.1) \qquad \int_{\theta_0+b}^{\theta+b} \hat{\theta}_0(x) f_0(x-\theta) dx = \theta - \theta_0 \int_{\theta}^{\theta_0+b} \hat{\theta}_0(x) f_0(x-\theta) dx$$

which is obtained from the condition

$$E_\theta \left[\hat{\theta}_0(X) \right] = \theta \quad \text{for} \quad \theta_0 < \theta \leq \theta_0 + (b-a).$$

It is seen that we can obtain a solution of the equation (1.3.1) by differentiating w.r.t. θ. Repeating the similar process, we can get $\hat{\theta}_0(x)$ for all values of x.

Example 1.3.2. Let X be a random variable according to a triangular distribution with a density function

$$f(x, \theta) = \begin{cases} 1 - |x - \theta| & \text{for} \quad |x - \theta| < 1 ; \\ 0 & \text{for} \quad |x - \theta| \geq 1. \end{cases}$$

Then it is shown that

$$\inf_{\hat{\theta}:\text{unbiased}} V_{\theta_0}\left(\hat{\theta}(X) \right) = 0,$$

but the lower bound is not attained. Fix $\theta_0 = 0$. Let us define an estimator of the type

$$\hat{\theta}_\varepsilon(x) = \begin{cases} \dfrac{c}{f_0(x)} = \dfrac{c}{1-x} & \text{for} \quad 0 \leq x < 1 - \varepsilon ; \\ -\dfrac{c}{f_0(x)} = -\dfrac{c}{1+x} & \text{for} \quad -1 + \varepsilon < x < 0 ; \\ 0 & \text{for} \quad 1 - \varepsilon \leq x < 1, \ -1 < x \leq -1 + \varepsilon, \end{cases}$$

where $0 < \varepsilon < 1$ and c is some constant. By the similar process as with $\hat{\theta}_0$ in Example 1.3.1 we can so construct $\hat{\theta}_\varepsilon(x)$ for all values of x that it is unbiased. And we can prove that

$$V_0(\hat{\theta}_\varepsilon) \to 0 \qquad \text{as} \quad \varepsilon \to 0$$

which implies that $\displaystyle\inf_{\hat{\theta}:\text{unbiased}} V_0(\hat{\theta}) = 0$. Note that here we can not let $\varepsilon = 0$, because we can not construct an unbiased estimator outside the interval $(-1, 1)$ of x.

Morimoto and Sibuya (1967) discussed the estimation of so-called selection parameter. Let $X_1, \ldots, X_n$ be i.i.d. random variables with the density function (with respect to a non-atomic measure μ) of the type

$$f(x, \theta) = \begin{cases} c(\theta) f(x) & \text{for} \quad x \in A(\theta), \\ 0 & \text{for} \quad x \notin A(\theta), \end{cases}$$

where $f(x)$ is a function independent of the real parameter θ. Let us define $T_1 = t_1(X_1, \ldots, X_n)$ and $T_2 = t_2(X_1, \ldots, X_n)$ by

$$T_1 = \inf \{\theta \,|\, X_i \in A(\theta) \quad (i = 1, \ldots, n)\}$$

and

$$T_2 = \sup \{\theta \,|\, X_i \in A(\theta) \quad (i = 1, \ldots, n)\}.$$

Obviously, the pair (T_1, T_2) $(T_1 \leq T_2)$ is sufficient for θ (with proper measurability condition etc.). Then assuming that the distribution of (T_1, T_2) is absolutely continuous w.r.t. the Lebesgue measure, the density function is given as

$$f^*(t_1, t_2, \theta) = \begin{cases} c(\theta)g(t_1, t_2) & \text{for} \quad a(\theta) < t_1 \leq t_2 < b(\theta), \\ 0, & \text{otherwise.} \end{cases}$$

Generally, (T_1, T_2) is complete if $\theta \geq c$ with $a(\theta) < b(\theta)$ and $a(\theta)$ is monotone decreasing and $b(\theta)$ is monotone increasing, and any estimable function $\eta(\theta)$ admits an UMV estimator. And if both $a(\theta)$ and $b(\theta)$ is monotone increasing, (T_1, T_2) is not complete, and under some regularity conditions there exists an unbiased estimator $\hat{\theta}_0$ of θ which has zero variance at any specified value θ_0.

Example 1.3.3. Let $X_1, \ldots, X_n$ be i.i.d. random variables with a density function

$$f(x, \theta) = \begin{cases} f(x)/F(\theta), & 0 \leq x \leq \theta, \\ 0, & \theta < x, \end{cases}$$

where $f(x) > 0$ a.e. and $F(\theta) = \int_0^\theta f(x)dx < \infty$. We consider the estimation of the selection parameter θ of the family $\{f(x, \theta) \,|\, 0 < \theta < \infty\}$. A minimal sufficient statistic for θ is $T = \max(X_1, \ldots, X_n)$ and its density function is given by

$$g(t, \theta) = \begin{cases} g(t)/G(\theta), & 0 \leq t \leq \theta \,; \\ 0, & \theta < t, \end{cases}$$

where $g(t) = nF^{n-1}(t)f(t)$ and $G(\theta) = F^n(\theta)$. T is complete, and if $\eta(\theta)$ is an absolutely continuous function defined on $(0, \infty)$ and if $\lim_{\varepsilon \to +0} \eta(\varepsilon)G(\varepsilon) = 0$, the UMVU estimator of $\eta(\theta)$ is given by

$$\phi(t) = \frac{1}{g(t)} \{\eta(t)G(t)\}' = \eta(t) + \eta'(t)\frac{G(t)}{g(t)} = \eta(t) + \eta'(t)\frac{F(t)}{nf(t)}.$$

The estimator $\phi(T)$ has the variance

$$V_\theta(\phi(T)) = \frac{1}{G(\theta)} \int_0^\theta \frac{\{\eta'(t)G(t)\}^2}{g(t)} dt = \frac{1}{nF^n(\theta)} \int_0^\theta \{\eta'(t)\}^2 \frac{F^{n+1}(t)}{f(t)} dt.$$

Example 1.3.4. Let $X_1, \ldots, X_n$ be i.i.d. random variables with a density function

$$f(x, \theta) = \begin{cases} f(x)/F(\theta), & \theta \leq x \leq b(\theta), \\ 0, & \text{otherwise}, \end{cases}$$

where $f(x) > 0$ a.e. and $F(\theta) = \int_\theta^{b(\theta)} f(x)dx < \infty$. A minimal sufficient statistic for $(\theta, b(\theta))$ is the pair of $T_1 = \min(X_1, \ldots, X_n)$ and $T_2 = \max(X_1, \ldots, X_n)$. Then the family of its densities $g_\theta(t) = g(t_1, t_2)/G(\theta)$ with

$$g(t_1, t_2) = n(n-1) \left\{ \int_{t_1}^{t_2} f(x)dx \right\}^{n-2} f(t_1)f(t_2)$$

and

$$G(\theta) = \left\{ \int_\theta^{b(\theta)} f(x)dx \right\}^n,$$

is not complete. Assume that $b(\theta)(> 0)$ is a strictly increasing and a.e. differentiable function. It will be possible to construct an unbiased estimator $\phi(X)$ of $\eta(\theta)$, based on a single observation X, with zero variance at a given value θ_0 of the parameter. Obviously, an estimator with this property must satisfy

$$\phi(x) = \eta(\theta_0) \qquad \text{a.e.,} \quad \theta_0 \leq x \leq b(\theta_0).$$

And from its unbiasedness

$$E_\theta[\phi(X)] = \int_\theta^{b(\theta)} \phi(x) \frac{f(x)}{F(\theta)} dx = \eta(\theta) \qquad \text{for all } \theta,$$

which implies the relation

$$b'(x)\phi(b(x)) f(b(x)) - \phi(x)f(x) = [\eta(x)F(x)]'$$

for almost all x.

As a special case in the example let $b(\theta) = 2\theta$, $F(\theta) = \theta$, $f(x) = 1$ and $\eta(\theta) = \theta$. Then we have

(1.3.2) $2\phi(2x) - \phi(x) = 2x$

for all x. Letting $\theta_0 = 1$, we obtain $\phi(x) = 1$ for $1 \leq x < 2$. From (1.3.2) we have

$$\phi(x) = \frac{1}{2}(x + 1) \qquad \text{for } 2 \leq x < 4,$$

$$\phi(x) = \frac{1}{2}\left(\frac{5}{4}x + \frac{1}{2}\right) \qquad \text{for } 4 \leq x < 8,$$

$$\phi(x) = \frac{1}{8}\left(\frac{21}{4}x + 1\right) \qquad \text{for } 8 \leq x < 16,$$

$\cdots$.

Similarly we obtain from (1.3.2)

$$\phi(x) = 2 - 2x \qquad \text{for} \quad \frac{1}{2} \le x < 1,$$

$$\phi(x) = 4 - 10x \qquad \text{for} \quad \frac{1}{4} \le x < \frac{1}{2},$$

$$\phi(x) = 8 - 42x \qquad \text{for} \quad \frac{1}{8} \le x < \frac{1}{4},$$

$$\cdots.$$

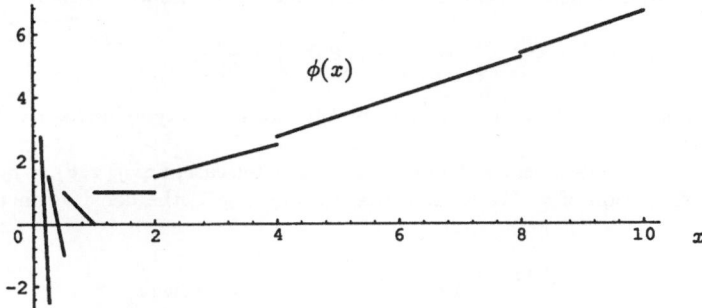

Figure 1.3.1. The unbiased estimator $\phi(X)$ given in the above way in the case of uniform distribution.

In other cases of non-constant support, we have bounds for the variance of unbiased estimators. Even when the support depends on θ, we sometimes have the Cramér-Rao type bounds. Suppose that a random variable X is distributed with the density function $f_0(x - \theta)$ with respect to the Lebesgue measure where $f_0(x)$ is defined as

$$f_0(x) = \begin{cases} c(1 - x^2)^2, & |x| \le 1, \\ 0, & \text{otherwise,} \end{cases}$$

where c is some constant. Then for any unbiased estimator $\hat{\theta}(x)$ we have

$$\int_{\theta-1}^{\theta+1} \hat{\theta}(x) f_0(x - \theta) dx = \theta.$$

By differentiating with respect to θ we have

$$\hat{\theta}(\theta + 1) f_0(1) - \hat{\theta}(\theta - 1) f_0(-1) - \int_{\theta-1}^{\theta+1} \hat{\theta}(x) f_0'(x - \theta) dx = 1.$$

Here the proof of differentiability is needed, but we omit its detailed discussion. Since $f_0(1) = f_0(-1) = 0$ we have

$$\int_{\theta-1}^{\theta+1} \hat{\theta}(x) f_0'(x - \theta) dx = -1.$$

By putting $\theta = 0$ and applying the Cauchy-Schwarz inequality we have

$$\left\{ \int_{-1}^{1} \hat{\theta}^2(x) f_0(x) dx \right\} \left\{ \int_{-1}^{1} \left(\frac{f_0'(x)}{f_0(x)} \right)^2 f_0(x) dx \right\} \geq \left\{ \int_{-1}^{1} \hat{\theta}(x) f_0'(x) dx \right\}^2 = 1.$$

Hence we obtain

$$V_0\left(\hat{\theta}(X) \right) \geq 1 \Big/ \int_{-1}^{1} \frac{(f_0'(x))^2}{f_0(x)} dx = \frac{3}{32c}.$$

What is remarkable here is that the bound is sharp, that is, we can show

$$\inf_{\hat{\theta}\,:\,\text{unbiased}} V_0\left(\hat{\theta}(X) \right) = \frac{3}{32c}.$$

Such an argument can be generalized to the Bhattacharyya type inequality (see Sections 2.2 to 2.4).

Example 1.3.5. We consider the location parameter case, i.e. $f(x, \theta) = f_0(x - \theta)$, and unbiased estimation of θ. We assume that for any $p \geq 1$, the density function $f_0(x)$ is given by

$$f_0(x) = \begin{cases} c(1 - x^2)^{p-1} & \text{if } |x| < 1, \\ 0 & \text{otherwise,} \end{cases}$$

where $c = 1/B(1/2, p)$ with $B(\alpha, \beta) = \int_0^1 x^{\alpha-1}(1-x)^{\beta-1} dx$ $(\alpha > 0, \beta > 0)$.

Case (i). Let $p = 1$. Then the distribution is uniform, and it is easy to check that

$$\min_{\hat{\theta}\,:\,\text{unbiased}} V_{\theta_0}(\hat{\theta}) = 0$$

for any specific value θ_0 (Takeuchi, 1961b). Indeed, if $X_1, \ldots, X_n$ are independently, identically and uniformly distributed random variables on the interval $(\theta - 1, \theta + 1)$, then the estimator $\hat{\theta}^*(X_1, \ldots, X_n) = (1/n)\sum_{i=1}^{n}[X_i - \theta_0 + 1] + \theta_0$, where $[a]$ denotes the maximum integer not exceeding a.

Case (ii). Let $p = 2$. In this case, it is easy to check that the amount of Fisher information is infinite, i.e.

$$\int_{-\infty}^{\infty} \left\{ \frac{f_0'(x)}{f_0(x)} \right\}^2 f_0(x) dx = \infty.$$

Then it is shown that

$$\inf_{\hat{\theta}\,:\,\text{unbiased}} V_{\theta_0}(\hat{\theta}) = 0$$

for any specified value θ_0. Before proceeding on to the next cases, we note the following: Let Λ be a $k \times k$ non-negative definite matrix whose elements are

$$(1.3.3) \qquad \lambda_{ij} = \int_{\theta-1}^{\theta+1} \frac{1}{f_0(x-\theta)} \frac{\partial^i f_0(x-\theta)}{\partial \theta^i} \frac{\partial^j f_0(x-\theta)}{\partial \theta^j} dx$$

$$= \int_{-1}^{1} (-1)^{i+j} \frac{1}{f_0(x)} f_0^{(i)}(x) f_0^{(j)}(x) dx, \qquad i, j = 1 \ldots, k.$$

If $k < p/2$, $\lambda_{ii}(i = 1,\ldots,k)$ are finite since $\int_{-1}^{1}(1 - x^2)^{p-2k-1}dx < \infty$. Then the Bhattacharyya bound for the variance of unbiased estimator $\hat{\theta}(X)$ of θ is given by

$$(1.3.4) \qquad V_{\theta_0}(\hat{\theta}) \geq (1,0\ldots,0)\Lambda^{-1}(1,0\ldots,0)'$$

for any specified value θ_0 (see Section 2.2). We obtain for $|x| < 1$,

$$(1.3.5) \quad \begin{aligned} f_0^{(1)}(x) &= -2c(p - 1)x(1 - x^2)^{p-2}, \\ f_0^{(2)}(x) &= -2c(p - 1)\left\{(1 - x^2)^{p-2} - 2(p - 2)x^2(1 - x^2)^{p-3}\right\}, \\ f_0^{(3)}(x) &= 4c(p - 1)(p - 2)\left\{3x(1 - x^2)^{p-3} - 2(p - 2)x^3(1 - x^2)^{p-4}\right\}, \\ &\cdots \end{aligned}$$

If $i+j$ is an odd number, it follows by (1.3.3) and (1.3.5) that $\lambda_{ij} = 0$ since $f_0^{(i)}(x)f_0^{(j)}(x)$ is an odd function. For $i + j$ even cases, λ_{ij} can be expressed in terms of beta functions.
Case (iii). Let $p = 3, 4, 5, 6$. Then it is shown that

$$\inf_{\hat{\theta}\,:\,\text{unbiased}} V_{\theta_0}(\hat{\theta}) = \frac{1}{\lambda_{11}}$$

for any specific value θ_0.
Case (iv). Let $p = 7$. In this case, we see that $k = 3$. Using (1.3.3), (1.3.4) and (1.3.5), we obtain

$$V_{\theta_0} \geq \frac{1}{|\Lambda|}\begin{vmatrix} \lambda_{22} & 0 \\ 0 & \lambda_{33} \end{vmatrix} = \left[\lambda_{11}\left(1 - \frac{\lambda_{13}^2}{\lambda_{11}\lambda_{33}}\right)\right]^{-1}$$

where

$$\Lambda = \begin{pmatrix} \lambda_{11} & 0 & \lambda_{13} \\ 0 & \lambda_{22} & 0 \\ \lambda_{13} & 0 & \lambda_{33} \end{pmatrix}.$$

Here again $\inf_{\hat{\theta}\,:\,\text{unbiased}} V_{\theta_0}(\hat{\theta}) = \left\{\lambda_{11}\left(1 - \frac{\lambda_{13}^2}{\lambda_{11}\lambda_{33}}\right)\right\}^{-1}$ for any specific θ_0, i.e. we have a sharp bound.
Case (v). For $p \geq 8$, we can continue in a similar manner by choosing $k = [(p - 1)/2]$, where $[s]$ denotes the largest integer less than or equal to s.

The asymptotics in similar situations to this section is given by Akahira (1975a, 1976, 1982), Akahira and Takeuchi (1981, 1991), Antoch (1984), Boente and Fraiman (1988), Jurečková (1981), Polfeldt (1970a, b), Smith (1985, 1989), Weiss (1979), Woodroofe (1972, 1974) and others.

1.4. DISCRETE PARAMETER SET

In some cases the parameter set itself can be discrete. Then the parameter set is either finite or countable. The case of finite parameter set can be generalized to the case of finite rank, which is defined as follows. The class $\{P_\omega|\ \omega \in \Omega\}$ of probability distributions is said to be of finite rank m if there exist $\omega_1, \ldots, \omega_m$ in Ω such that for any $\omega \in \Omega$ there exists a set of constants $c_1(\omega), \ldots, c_m(\omega)$ satisfying

$$P_\omega = c_1(\omega)P_{\omega_1} + \cdots + c_m(\omega)P_{\omega_m} \qquad \text{for all} \quad \omega \in \Omega$$

and $P_{\omega_1}, \ldots, P_{\omega_m}$ are linearly independent in the sense that $a_1 P_{\omega_1} + \cdots + a_m P_{\omega_m} \equiv 0$ implies $a_1 = \cdots = a_m = 0$. Then for any real-valued function $\theta = \theta(\omega)$ of ω, an unbiased estimator exists if and only if

$$\theta(\omega) = c_1(\omega)\theta(\omega_1) + \cdots + c_m(\omega)\theta(\omega_m)$$

for all $\omega \in \Omega$. The class of all UMVU estimators (of any parameter) is equal to the class of functions measurable with respect to a finite subfield of $\mathcal{A}$. Indeed, we can show that if $\hat\theta$ is UMVU for θ then so is $\hat\theta^k$ for $k = 2, 3, \ldots$ (see, e.g. Rao (1952)). But it is obvious that there is no more than m linearly independent UMVU estimators, hence we must have

$$\hat\theta^r + a_1 \hat\theta^{r-1} + \cdots + a_{r-1}\hat\theta + a_r = 0$$

for some $r(\le m+1)$, which implies that $\hat\theta$ takes at most $m+1$ different values. It follows that all UMVU estimators must be measurable with respect to a finite subfield of $\mathcal{A}$ which generated by at most $m+1$ disjoint sets.

In the case of countable index set we have a LMVU estimator for any parameter point $\omega = \omega_0$ whose variance is infinity except for $\omega = \omega_0$, by the method of Stein (1950) (see Section 1.5).

Example 1.4.1 (Kojima et al., 1982). Let $X_1, \ldots, X_n$ be i.i.d. random variables according to a normal distribution with mean m and variance nd^2, where m is a integer-valued parameter. The sample mean $\bar{X} = \sum_{i=1}^n X_i/n$ is sufficient and its distribution is normal with mean m and variance d^2. Then the LMVU estimator at $m = 0$ is given by

$$f(\bar{X}) = \sum_{u=1}^\infty (-1)^{u+1} \left\{ \exp\left(u(u-1)/2d^2\right) \left(\exp(u/d^2) - 1\right) \exp\left(u^2/2d^2\right) \right\}^{-1}$$
$$\cdot \left\{ \exp\left(u\bar{X}/d^2\right) - \exp\left(-u\bar{X}/d^2\right) \right\}.$$

Its local minimum variance at $m = 0$ is

$$V(f) = 2 \sum_{u=1}^\infty (-1)^{u+1} u \left\{ \exp(-u(u-1)/2d^2)/(\exp(u/d^2) - 1) \right\}.$$

The LMVU estimator at $m = 0$ has infinite variance at all $m \neq 0$.

When the sample space $\mathcal{X}$ itself is a finite set of size N, then the probability distribution P_ω can be considered to be a N-dimensional real vector, and if there exist N linearly independent P_{ω_i} ($i = 1, \ldots, N$) then the sample X is complete and any parameter with unbiased estimators has a unique unbiased estimator which is UMV unbiased by the definition.

Example 1.4.2. Let X be distributed according to a hypergeometric distribution

$$P\{X = x\} = \frac{{}_M C_x \cdot {}_{N-M} C_{n-x}}{{}_N C_n} \quad \text{for} \quad x = 0, 1, \ldots, n,$$

where $n < M$ and $2n < N$. Let M be the unknown parameter and the possible values of M be $n+1, \ldots, N-n$. Then X is complete.

1.5. DISCONTINUOUS AND NON-DIFFERENTIABLE DENSITY

The approach adopted by Barankin (1949) and Stein (1950) can be briefly sketched as follows. Let X be a random variable according to a density function $f(x, \theta)$ with respect to a σ-finite measure μ, where $x \in \mathcal{X}$ and $\theta \in \Theta$. Fix θ_0 in Θ and consider the L_2-space $\mathcal{M}$ with a norm $\| \cdot \|$ defined by

$$\|h\| := \left\{ \int_{\mathcal{X}} |h(x)|^2 f(x, \theta_0) d\mu(x) \right\}^{1/2}.$$

Let $\mathcal{M}_0$ be a subspace of $\mathcal{M}$ which is spanned by the functions

$$\lambda_\theta(x) := f(x, \theta)/f(x, \theta_0),$$

assuming $\lambda_\theta \in \mathcal{M}$. Then, for an unbiased estimator $\gamma(X)$ of $g(\theta)$, we have

$$< \gamma, \lambda_\theta >= g(\theta),$$

where $< \cdot, \cdot >$ designates the inner product. Hence, for any unbiased estimator γ, we have its projection γ^* onto $\mathcal{M}_0$ such that

$$< \gamma^*, \lambda_\theta >=< \gamma, \lambda_\theta > \quad \text{for all} \quad \theta \in \Theta$$

and $\|\gamma^*\| \leq \|\gamma\|$. Such γ^* is unique, hence is the LMVU estimator of $g(\theta)$ at $\theta = \theta_0$. Now suppose that the density function $f(x, \theta)$ is not continuous with respect to θ while the support $A = \{x|\ f(x, \theta) > 0\}$ can be defined to be independent of θ. Note that this condition does not affect the existence of a LMVU estimator, because it can be shown that for any real $\gamma(\theta)$ which has an unbiased estimator, a LMVU estimator at $\theta = \theta_0$ exists if

$$\int_A \frac{\{f(x, \theta)\}^2}{f(x, \theta_0)} d\mu < \infty \quad \text{for all} \quad \theta$$

(see Barankin (1949) and Stein (1950)). But the LMVU estimator sometimes behaves very strangely (see Section 2.5).

Example 1.5.1. Let $X_1, \ldots, X_n$ be i.i.d. random variables with the following density

$$f(x, \theta) = \begin{cases} p, & 0 \leq x \leq \theta \quad \text{and} \quad \theta + 1 \leq x \leq 2, \\ q, & \theta < x < \theta + 1, \\ 0, & \text{otherwise}, \end{cases}$$

where p and q with $0 < p < q$ and $p + q = 1$ are fixed constants. Since the support of $f(x, \theta)$ is the interval $[0, 2]$, in the subsequent discussion it is enough to consider it as the domain of x. The parameter θ has the range $0 \leq \theta \leq 1$. Then the LMVU estimator of θ at $\theta = \theta_0$ has the form

$$\hat{\theta}_0(x_1, \ldots, x_n) = \int_0^1 \prod_{i=1}^n \frac{f(x_i, \eta)}{f(x_i, \theta_0)} dG_n(\eta),$$

where G_n is some signed measure over the closed interval $[0, 1]$ (see Stein (1950)).

For $n \geq 2$, we take a signed measure G_n^* over $[0, 1]$ satisfying

$$\frac{dG_n^*}{d\eta} = \left(1 - \frac{1}{n}\right) \frac{1}{(1 + c|\eta - \theta_0|)^n} \, \text{sgn}(\eta - \theta_0) \qquad \text{for} \quad 0 < \eta < \theta_0, \, \theta_0 < \eta < 1 \, ;$$

$$G_n^*(\{0\}) = -\frac{1}{cn(1 + c\theta_0)^{n-1}} \, ;$$

$$G_n^*(\{1\}) = \frac{1}{cn(1 + c(1 - \theta_0))^{n-1}} \, ;$$

$$G_n^*(\{\theta_0\}) = \theta_0.$$

It is shown that

$$\hat{\theta}_0^*(x_1, \ldots, x_n) = \int_0^1 \prod_{i=1}^n \frac{f(x_i, \eta)}{f(x_i, \theta_0)} dG_n^*(\eta)$$

is a LMVU estimator of θ which has the following variance at $\theta = \theta_0$:

$$V_{\theta_0}\left(\hat{\theta}_n^*\right) = \begin{cases} \dfrac{1}{2c^2} \log\{(1 + c\theta_0)(1 + c(1 - \theta_0))\} & \text{for} \quad n = 2 \, ; \\[3mm] \dfrac{1}{c^2 n(n-2)} \left\{2 - \dfrac{1}{(1 + c\theta_0)^{n-2}} - \dfrac{1}{(1 + c(1 - \theta_0))^{n-2}}\right\} & \text{for} \quad n > 2. \end{cases}$$

In the case of discontinuous or non-differentiable density functions a lower bound for the variance of unbiased estimators of a real parameter θ was obtained by Chapman and Robbins (1951), which is given by

$$V_{\theta_0}\left(\hat{\theta}(X)\right) \geq \sup_\eta \frac{(\eta - \theta_0)^2}{\int_{\mathcal{X}} \left\{\dfrac{f(x, \eta)}{f(x, \theta_0)} - 1\right\}^2 d\mu}.$$

It is expressed as

$$V_{\theta_0}\left(\hat{\theta}(X)\right) \geq \left[\inf_\eta \frac{K_{\theta_0}(\eta)}{(\eta - \theta_0)^2}\right]^{-1},$$

where

$$K_{\theta_0}(\eta) = \int_{\mathcal{X}} \frac{\{f(x, \eta) - f(x, \theta_0)\}^2}{f(x, \theta_0)} d\mu.$$

When $X_1, \ldots, X_n$ are i.i.d. random variables with density $f(x, \theta)$,

$$(1.5.1) \qquad V_{\theta_0}\left(\hat{\theta}(X_1, \ldots, X_n)\right) \geq \left[\inf_{\eta} \frac{\{K_{\theta_0}(\eta) + 1\}^n - 1}{(\eta - \theta_0)^2}\right]^{-1}.$$

In the regular cases we have

$$K_{\theta_0}(\eta) = I(\theta)(\eta - \theta_0)^2 + o((\eta - \theta_0)^2)$$

and the right-hand side of (1.5.1) is reduced to the Cramér-Rao bound when $\eta - \theta_0$ is of the order $1/\sqrt{n}$. In the non-regular cases we sometimes have

$$K_{\theta_0}(\eta) = I(\theta)|\eta - \theta_0|^\alpha + o(|\eta - \theta_0|^\alpha),$$

where $\alpha > 2$.

1.6. NON SQUARE-INTEGRABLE LIKELIHOOD RATIO

There are cases when the support $S = \{x \mid f(x, \theta) > 0\}$ is independent of θ, but $\int_S \{f(x, \theta)\}^2 / f(x, \theta_0) d\mu$ is infinite for some $\theta \in \Theta$. Assume that $\Theta = \{\theta_0, \theta_1, \ldots, \theta_p\}$ is a finite set. Then we consider to minimize

$$\int_S \{\hat{\theta}(x)\}^2 f(x, \theta_0) d\mu$$

under the unbiasedness condition : $E_{\theta_i}(\hat{\theta}) = \gamma(\theta_i) = \gamma_i$ (say) $(i = 1, \ldots, k)$. Assume that $f(x, \theta_i)$ $(i = 1, \ldots, p)$ are linearly independent,

$$\int_S \frac{\{f(x, \theta_i)\}^2}{f(x, \theta_0)} d\mu < \infty \qquad (i = 1, \ldots, k \ ; \ k \leq p),$$

and moreover

$$\int_S \frac{\{\sum_{i=1}^p c_i f(x, \theta_i)\}^2}{f(x, \theta_0)} d\mu < \infty$$

implies

$$c_{k+1} = \cdots = c_p = 0.$$

Let Λ be a $k \times k$ non-negative definite matrix whose elements are

$$\lambda_{ij} = \int_S \frac{f(x, \theta_i) f(x, \theta_j)}{f(x, \theta_0)} d\mu \qquad (i, j = 1, \ldots, k).$$

Then it is shown by Takeuchi and Akahira (1985) that under the condition

$$\inf_{\hat{\theta} : \text{unbiased}} \int_S \left\{\hat{\theta}(x)\right\}^2 f(x, \theta_0) d\mu = \gamma(k)' \gamma(k)$$

holds, where $\gamma(k) = (\gamma_1, \ldots, \gamma_k)'$ is given in the above (see Section 2.1). Therefore in this case the lower bound for the variance of unbiased estimators can be obtained by simply disregarding of $(\theta_{k+1}, \ldots, \theta_p)$, however, it should be noted that the lower bound is not generally attained.

1.7. ASYMPTOTIC THEORY FOR NON-REGULAR CASES

We recollect that, in the regular case when $X_1, \ldots, X_n$ are i.i.d. random variables with a density $f(x, \theta)$ which is differentiable with respect to θ up to the required degree, it is shown that (i) the maximum order of consistency is $\sqrt{n}$, (ii) an estimator $\hat{\theta}_n$ is asymptotically efficient if $\sqrt{n}(\hat{\theta}_n - \theta)$ is asymptotically normally distributed with mean 0 and variance $1/I(\theta)$, where $I(\theta)$ is the amount of Fisher information on $f(x, \theta)$, (iii) the maximum likelihood estimator (MLE) is asymptotically efficient, (iv) first order efficiency implies second order efficiency, (v) third order asymptotically efficient estimators do not usually exist but the modified MLE is third order asymptotically efficient either (a) symmetrically or (b) among the class **D** of estimators (see, e.g. Akahira and Takeuchi (1981)).

In non-regular situations, all of the above statements fail to apply in various cases. The maximum order of consistency can be $n^{1/\alpha}$ ($0 < \alpha < 2$) and $\sqrt{n \log n}$ which was discussed in our previous monograph. In this volume we shall give somewhat different approach to the problem on the maximum order of consistency (see Section 3.5). Once the maximum order of consistency being established, the existence of asymptotically efficient estimators will be discussed, and the answer is usually negative, which was also shown in our previous monograph. Furthermore, if the second order differentiability of the density function does not hold, the first order efficiency does not necessarily mean the second order efficiency, as was exemplified by the case of double exponential distributions (also in Akahira and Takeuchi (1981)). We shall further discuss this phenomenon into more detail (see Chapter 4).

1.8. ASYMPTOTIC BAYES POSTERIOR DISTRIBUTION OF THE PARAMETER IN NON-REGULAR CASES

It has long been known that, in regular cases, the Bayes posterior distribution w.r.t. any smooth prior density approaches asymptotically normal with mean $\hat{\theta}_{ML}$ and variance $1/(nI(\hat{\theta}_{ML}))$ (see, e.g. Lindley (1953)), where $\hat{\theta}_{ML}$ denotes the MLE of θ. More precisely, when n is large, the posterior distribution of $\sqrt{n}(\theta - \hat{\theta}_{ML})$ is constant (i.e. normal with mean 0 and variance $1/I(\hat{\theta}_{ML})$) in the first order, and has only a location shift in the second order (i.e. of the order $n^{-1/2}$), This fact explains why the Bayes estimator (w.r.t. squared or any other symmetric loss) is asymptotically equivalent to the modified MLE up to the second (or in some cases up to the third) order. In other words, the Bayes estimator becomes asymptotically independent of the prior density or the loss function. But in non-regular cases, the above statements do not hold. The posterior density varies from sample to sample, even when it become independent of the prior density as n is large.

Consider the simplest case of the uniform distribution. Suppose that $X_1, \ldots, X_n$ are i.i.d. random variables with a uniform distribution over the interval $(\theta - (1/2), \theta + (1/2))$. If a prior density is $\pi(\theta)$, then the posterior density is given by

$$\begin{cases} \pi(\theta) \Big/ \displaystyle\int_{\underline{\theta}}^{\overline{\theta}} \pi(\theta) d\theta & \text{for} \quad \underline{\theta} < \theta < \overline{\theta}, \\ 0 & \text{otherwise}, \end{cases}$$

where $\underline{\theta} = \max_{1 \leq i \leq n} X_i - (1/2)$, $\overline{\theta} = \min_{1 \leq i \leq n} X_i + (1/2)$. Considering the fact that $\overline{\theta} - \underline{\theta} = 1 - (\max_{1 \leq i \leq n} X_i - \min_{1 \leq i \leq n} X_i)$ is of order $1/n$, we can show that the posterior distribution of $n \left(\theta - \hat{\theta}\right)$ is uniform over the interval $\left(-n \left(\overline{\theta} - \underline{\theta}\right)/2, n \left(\overline{\theta} - \underline{\theta}\right)/2\right)$, where $\hat{\theta} = \left(\overline{\theta} + \underline{\theta}\right)/2$, provided that $\pi(\theta)$ is smooth. Then it is shown that the distribution of $n \left(\overline{\theta} - \underline{\theta}\right)$ approaches to the exponential one, but not to the constant. This fact shows that the asymptotic posterior density depends not only on $\hat{\theta}$ but also on $R := \max_{1 \leq i \leq n} X_i - \min_{1 \leq i \leq n} X_i$ which means that $\hat{\theta}$ is not asymptotically sufficient, and which suggests also that $\hat{\theta}$ will not be asymptotically efficient.

A slightly different example in the case that $X_1, \ldots, X_n$ are i.i.d. random variables with the density

$$f(x, \theta) = \begin{cases} c \exp\left\{-\dfrac{1}{2}(x - \theta)^2\right\} & \text{for} \quad |x - \theta| < 1, \\ 0 & \text{otherwise}. \end{cases}$$

If a prior density is $\pi(\theta)$, then the posterior density is given by

$$\begin{cases} \left[\pi(\theta) \exp\left\{-\dfrac{n}{2}(\bar{x} - \theta)^2\right\}\right] \Big/ \displaystyle\int_{\underline{\theta}}^{\overline{\theta}} \pi(\theta) \exp\left\{-\dfrac{n}{2}(\bar{x} - \theta)^2\right\} d\theta & \text{for} \quad \underline{\theta} < \theta < \overline{\theta}, \\ 0 & \text{otherwise}, \end{cases}$$

where $\bar{x} = \sum_{i=1}^{n} x_i / n$, $\underline{\theta} = \max_{1 \leq i \leq n} X_i - 1$ and $\overline{\theta} = \min_{1 \leq i \leq n} X_i + 1$. We can also observe that the posterior distribution of $n(\theta - \hat{\theta})$ is almost uniform, but with a slight slant depending on the sign of $\bar{X} - \hat{\theta}$, where $\hat{\theta} = \left(\overline{\theta} + \underline{\theta}\right)/2$. From this it is derived that the Bayes mode, which is equivalent to the MLE, is either $\underline{\theta}$ or $\overline{\theta}$, but is clearly inferior to $\hat{\theta}$ for most of usual loss functions. This explains why in this case the MLE is asymptotically *inadmissible*.

As another, more difficult case, consider the case when the density $f(x, \theta)$ is given in Example 1.5.1. If a prior density is $\pi(\theta)$, then the posterior density is proportional to

$$\pi(\theta) \left(\frac{q}{p}\right)^{\sum_{i=1}^{n} \chi_{\theta}(x_i)},$$

for $0 \leq x_i \leq 2$ $(i = 1, \ldots, n)$, where

$$\chi_\theta(x) = \begin{cases} 1 & \text{for} \quad \theta < x < \theta + 1, \\ 0 & \text{otherwise.} \end{cases}$$

We can also observe that $\sum_{i=1}^{n} \chi_\theta(X_i)$ is far from smooth, even if we can show that

$$\frac{1}{n} \sum_{i=1}^{n} \chi_\theta(X_i) \xrightarrow{P_{\theta_0}} E_{\theta_0}[\chi_\theta(X)] = q - (q-p)|\theta - \theta_0|$$

and, asymptotically, $n\left(\theta - \hat{\theta}_{ML}\right)$ is double exponentially distributed, where $\xrightarrow{P_{\theta_0}}$ means the convergence in probability P_{θ_0} as $n \to \infty$. In such cases, non-smoothness of the posterior distribution makes trouble with the MLE.

A more delicate situation is the case when the asymptotic posterior distribution is well behaved but not in the second order, as in the case of location parameter with double exponential distribution. It is shown that the posterior distribution is asymptotically normal with constant variance, but in the second order, the density is not smooth, which is related to the fact that in this case the MLE is first order efficient but not second order efficient (see Fisher (1925)).

1.9. OVERVIEW

In the subsequent chapters the following are given. In Chapter 2 the lower bound for the variance of unbiased estimators is discussed. In Section 2.1, minimizing $\int \{\hat{\theta}(x)\}^2 f(x) d\mu$ is discussed under the unbiasedness condition stated in Section 1.6 (Takeuchi and Akahira, 1985). In Section 2.2, the Bhattacharyya bound is generalized to non-regular cases when the support of the density depends on the parameter, while it is differentiable several times with respect to the parameter within the support (Akahira, Puri and Takeuchi, 1986). In Sections 2.3 and 2.4 we introduce the concept of one-directionality which includes both cases of location (and scale) parameter and selection parameter but also other cases, and establish some theorems for sharp lower bounds and for the existence of zero variance unbiased estimator for this class of non-regular distributions (Akahira and Takeuchi, 1987a). In Section 2.5 the exact forms of the locally minimum variance unbiased estimators and their variances are given in the case of a discontinuous density function (Akahira and Takeuchi, 1987b).

In Chapter 3, relevance of the concept of amount of information to estimator is discussed. In Sections 3.1 and 3.2 the locally minimum variance of unbiased estimators is shown to be zero in the case when the amount of Fisher information is infinity and some examples are also given (Akahira and Takeuchi, 1985). In Sections 3.3 and 3.4, an amount of information is defined, its properties are discussed and some examples are given. In Section 3.5 the order of consistency is discussed from the viewpoint of the information amount (Akahira, 1994a).

In Chapter 4 comparisons of estimators are made in the cases of the double exponential and uniform distributions. In Section 4.1 we obtain the (asymptotic) losses of informations of order statistics and related estimators and asymptotically compare them using their asymptotic distributions up to the second order in the double exponential case (Akahira and Takeuchi, 1990). In Section 4.3 the problem to estimate a common parameter for the pooled sample from the uniform distributions is discussed in the presence of nuisance parameters, and the maximum likelihood estimator (MLE) and others are compared and it is shown that the MLE based on the pooled sample is not (asymptotically) efficient (Akahira and Takeuchi, 1985). In Section 4.4 we asymptotically compare the MLE, a weighted median, a weighted mean and others of a common parameter for the pooled sample from the double exponential distributions up to the second order, i.e. the order $n^{-1/2}$ (Akahira, 1984).

In Chapter 5 the problem is discussed to estimate an unknown real-valued common parameter θ based on the pooled samples for uniform and double exponential distributions (Akahira, 1984 ; Akahira and Takeuchi, 1985). In Sections 5.1 and 5.2 the MLE and others of the common parameter are compared for the samples of a fixed size and a large one in the uniform case. In Sections 5.3 and 5.4 the second order asymptotic comparison of estimators is discussed in the double exponential case.

In Chapter 6 the estimation problem of location parameter is discussed for the two-sided Weibull type distribution (Akahira and Takeuchi, 1989). In Sections 6.1 to 6.3 the higher order asymptotic bound for the distribution of second order asymptotically median unbiased estimators is obtained, and the higher order asymptotic distribution of the MLE is also given. In Section 6.4 the amount of the loss of asymptotic information of the MLE is given.

In Chapter 7 the second order asymptotic optimality of estimators is discussed for a family of truncated distributions (Akahira, 1988b, 1991a). In Sections 7.1 and 7.2 the second order asymptotic density of the generalized Bayes estimator is given. In Sections 7.3 and 7.4 the upper bound for the concentration probability around the true parameter for second order asymptotically median unbiased estimators is obtained and some examples are also given.

In Supplement, the order of consistency depending on the parameter is discussed (Akahira and Takeuchi, 1986). Indeed, such an order arises in the explosive case $|\theta| > 1$ in estimating a parameter θ in the first order autoregressive process. And also the bound for the asymptotic distribution of asymptotically median unbiased estimators of θ of the model in the unstable case $|\theta| = 1$ and the explosive case $|\theta| > 1$ is discussed.

CHAPTER 2

LOWER BOUND FOR THE VARIANCE
OF UNBIASED ESTIMATORS

2.1. MINIMUM VARIANCE

Let $(\mathcal{X}, \mathcal{B})$ be a sample space. We consider a family $\mathcal{P} = \{P_\theta \mid \theta \in \Theta\}$ of probability measures on $\mathcal{B}$ and let μ be a σ-finite measure on $\mathcal{B}$, where the index set Θ is called a parameter space. In minimum variance unbiased estimation theory, the locally best unbiased estimator $\hat{\gamma}^*$ of $\gamma = g(\theta)$ at $\theta = \theta_0$ is obtained by minimizing

$$\int_{\mathcal{X}} \{\hat{\gamma}(x)\}^2 \, dP_{\theta_0}(x)$$

under the condition that

(2.1.1) $$\int_{\mathcal{X}} \hat{\gamma}(x) dP_\theta(x) = g(\theta) \qquad \text{for all } \theta \in \Theta.$$

When the parameter space Θ is a finite set $\{\theta_0, \theta_1, \ldots, \theta_p\}$, the condition (2.1.1) is reduced to

$$\int_{\mathcal{X}} \hat{\gamma}(x) dP_{\theta_i}(x) = g(\theta_i) = c_i \quad (\text{say}) \quad (i = 0, 1, \ldots, p)$$

And it is easily derived that the minimizing solution $\hat{\gamma}^*$ is obtained as

$$\hat{\gamma}^*(x) = \sum_{i=0}^{p} \lambda_i f_i(x),$$

where

$$f_i(x) = \frac{dP_{\theta_i}}{dP_{\theta_0}}(x),$$

provided that

(A.2.1.1)   for each $i = 1, \ldots, p$, P_{θ_i} is absolutely continuous with respect to P_{θ_0},

(A.2.1.2) $$\int_{\mathcal{X}} \left(\frac{dP_{\theta_i}}{dP_{\theta_0}} \right)^2 dP_{\theta_0}(x) < \infty \qquad (i = 1, \ldots, p),$$

(A.2.1.3) $f_i(x)$ $(i = 1, \ldots, p)$ are linearly independent.

Then we have

$$E_{\theta_0}\left[\{\hat{\gamma}^*(X)\}^2\right] = \mathbf{c}'\Lambda^{-1}\mathbf{c},$$

where $\mathbf{c} = (c_1, \ldots, c_p)'$ and $\Lambda = \{\lambda_{ij}\}$ is a $n \times n$ non-negative definite matrix with

$$\lambda_{ij} = \int_{\mathcal{X}} f_i(x)f_j(x)dP_{\theta_0}(x) \quad (i, j = 1, \ldots, k).$$

The purpose of this section is to investigate the situation where for some i the first and/or the second condition does not hold. Remark that the third condition is obviously necessary. First we consider the case when we assume the first condition but not the second. To deal with the case we need a more subtle formulation of the condition in place of (A.2.1.2).

Let $f(x)$ and $f_i(x)$ $(i = 1, \ldots, p)$ be measurable functions. Then we consider to minimize

$$\int_{\mathcal{X}} \left\{\hat{\theta}(x)\right\}^2 f(x)d\mu$$

under the unbiasedness conditions :

$$\int_{\mathcal{X}} \hat{\theta}(x)f_i(x)d\mu = c_i \qquad (i = 1, \ldots, p)$$

and the condition (A.2.1.4) : $f_i(x)$ $(i = 1, \ldots, p)$ are linearly independent,

$$\int_{\mathcal{X}} \frac{\{f_i(x)\}^2}{f(x)}d\mu < \infty \qquad (i = 1, \ldots, k \; ; \; k \le p),$$

and

$$\int_{\mathcal{X}} \frac{\{\sum_{i=1}^{p} a_i f_i(x)\}^2}{f(x)}d\mu < \infty$$

implies

$$a_{k+1} = \cdots = a_p = 0.$$

Note that in the above f and f_i $(i = 1, \ldots, p)$ may not necessarily be density functions. In fact, in some applications we may take the derivatives of the density with respect to the parameter θ as f and obtain some generalization of Bhattacharyya inequality. Also generally f is equal to one of f_i's but it is not essential for the subsequent discussion. Let Λ_{11} be a $k \times k$ non-negative definite matrix whose elements are

$$\lambda_{ij} = \int_{\mathcal{X}} \frac{f_i(x)f_j(x)}{f(x)}d\mu \qquad (i, j = 1, \ldots, k).$$

Then we have the following.

Theorem 2.1.1. *Under the condition (A.2.1.4),*

$$\inf_{\hat{\theta}\,:\,\text{unbiased}} \int_{\mathcal{X}} \left\{\hat{\theta}(x)\right\}^2 f(x)d\mu = \mathbf{c}'_{(k)}\Lambda_{11}\mathbf{c}_{(k)}$$

holds, where $\mathbf{c}_{(k)} = (c_1, \ldots, c_k)'$ is given in the above. The equality is not usually attained.

Proof. We put

$$A_n = \left\{ x \;\middle|\; \max_{1 \le i \le p} \frac{|f_i(x)|}{f(x)} < n \right\} \qquad (n = 1, 2, \ldots).$$

We define the class C_n of estimators such that for any $\hat{\theta}_n \in C_n, \hat{\theta}_n(x) = 0$ for $x \notin A_n$. We also put

$$\lambda_{ij}^{(n)} = \int_{A_n} \frac{f_i(x) f_j(x)}{f(x)} d\mu \qquad (i, j = 1, \ldots, p)$$

and denote the matrix $\left(\lambda_{ij}^{(n)} \right)$ by Λ_n. Note that, for sufficiently large n, $f_i(x)$ ($i = 1, \ldots, p$) are linearly independent in A_n, hence Λ_n is nonsingular. Then we have

$$\min_{\hat{\theta}_n \in C_n \,:\, \text{unbiased}} E_\theta \left(\hat{\theta}_n^2 \right) = \mathbf{c}' \Lambda_n^{-1} \mathbf{c},$$

where $\mathbf{c} = (c_1, \ldots, c_p)'$ is given in the unbiasedness condition (A.2.1.4). On the other hand we may make a partition of the matrix Λ_n into blocks as

$$\Lambda_n = \begin{pmatrix} \Lambda_{11}^{(n)} & \Lambda_{12}^{(n)} \\ \Lambda_{21}^{(n)} & \Lambda_{22}^{(n)} \end{pmatrix},$$

where $\Lambda_{11}^{(n)}$ and $\Lambda_{22}^{(n)}$ are $k \times k$ and $(p-k) \times (p-k)$ matrices, respectively. Let $\mathbf{c}' = (\mathbf{c}_1', \mathbf{c}_2')$ with $\mathbf{c}_1 \in \mathbf{R}^k$ and $\mathbf{c}_2 \in \mathbf{R}^{p-k}$. Then the condition (A.2.1.4) implies that

$$(2.1.2) \qquad\qquad \lim_{n \to \infty} \mathbf{c}' \Lambda_n \mathbf{c} = \infty$$

for every vector $\mathbf{c}' = (\mathbf{c}_1', \mathbf{c}_2')$ such that $\mathbf{c}_2 \ne \mathbf{0}$. To prove the theorem it is necessary and sufficient to show that

$$\Lambda_n^{-1} \to \begin{pmatrix} \Lambda_{11}^{-1} & O \\ O & O \end{pmatrix} \qquad \text{as} \quad n \to \infty,$$

and since the matrix Λ_n is non-negative definite it is enough to show that

$$\left(\Lambda_{22}^{(n)} - \Lambda_{21}^{(n)} \Lambda_{11}^{(n)^{-1}} \Lambda_{12}^{(n)} \right)^{-1} \to O \qquad \text{as} \quad n \to \infty,$$

that is, the minimum characteristic root ρ_n of the matrix $\tilde{\Lambda}_n = \Lambda_{22}^{(n)} - \Lambda_{21}^{(n)} \Lambda_{11}^{(n)^{-1}} \Lambda_{12}^{(n)}$ diverges to infinity as n tends to be large. Assume otherwise, then we have a sequence $\{\mathbf{c}_n^*\}$ of vectors such that $\mathbf{c}_n^{*\prime} = \left(\mathbf{c}_{1n}^{*\prime}, \mathbf{c}_{2n}^{*\prime} \right)$ with $\|\mathbf{c}_{2n}^*\| = 1$ and

$$\mathbf{c}_n^{*\prime} \Lambda_n \mathbf{c}_n^* = \mathbf{c}_{2n}^{*\prime} \tilde{\Lambda}_n \mathbf{c}_{2n}^* = \rho_n$$

and $\rho_n \uparrow \rho^* (< \infty)$ as $n \to \infty$. Then there exists a subsequence $\{c_{n_j}^*\}$ of $\{c_n^*\}$ such that $c_{n_j}^{*\,'} = \left(c_{1n_j}^{*\,'}, c_{2n_j}^{*\,'} \right)$ with $\|c_{2n_j}^*\| = 1$ and

$$c_{n_j}^* \to c^* \qquad \text{as} \quad j \to \infty,$$

where $c^{*\,'} = (c_1^{*\,'}, c_2^{*\,'})$ with $\|c_2^*\| = 1$. We can show that

$$c^{*\,'} \Lambda_n c^* \to \rho^* \qquad \text{as} \quad n \to \infty,$$

which contradicts (2.1.2). Assume otherwise, since $c^{*\,'} \Lambda_n c^* \geq \rho_n$ for all n, there exist $\varepsilon > 0$ and n_0 such that

$$c^{*\,'} \Lambda_{n_0} c^* \geq \rho^* + \varepsilon.$$

Since Λ_n is monotone increasing in n in the sense that $c' \Lambda_n c$ is so for any vector c, it follows that for $n \geq n_0$

$$c_n^{*\,'} \Lambda_{n_0} c_n^* \leq c_n^{*\,'} \Lambda_n c_n^* \leq \rho^*.$$

Hence

$$\rho^* + \varepsilon \leq c^{*\,'} \Lambda_{n_0} c^* \leq \rho^*,$$

which is a contradiction. Thus we complete the proof.

Next we shall discuss the case when the first condition (A.2.1.1) does not hold true. We consider a family $\mathcal{P} = \{P_\theta : \theta \in \Theta\}$ of probability measures on $(\mathcal{X}, \mathcal{B})$, where $\Theta = \{\theta_1, \dots, \theta_p\}$. We assume that, for each $i = 1, \dots, p$, P_{θ_i} is absolutely continuous with respect to a σ-finite measure μ. For each $i = 1, \dots, p$ we denote $dP_{\theta_i}/d\mu$ by $f_i(x)$. Let $f(x)$ be one of $f_1(x), \dots, f_p(x)$. We denote by S the support of $f(x)$. Then we have the following problem : Minimize

$$\int_S \left\{ \hat{\theta}(x) \right\}^2 f(x) d\mu$$

under the unbiasedness condition

$$\int_{\mathcal{X}} \hat{\theta}(x) f_i(x) d\mu = \left(\int_S + \int_{S^c} \right) \hat{\theta}(x) f_i(x) d\mu = c_i \qquad (i = 1, \dots, p),$$

where $f_i(x)$ $(i = 1, \dots, p)$ are linearly independent. Assume that we have a condition similar to (A.2.1.4) above :

$$\int_S \frac{\{f_i(x)\}^2}{f(x)} d\mu < \infty \qquad (i = 1, \dots, k),$$

and

$$\int_S \frac{\{\sum_{i=1}^p c_i f_i(x)\}^2}{f(x)} d\mu < \infty$$

implies

$$c_{k+1} = \cdots = c_p = 0,$$

and further we assume that $f_i(x)$ $(i = 1, \ldots, p)$ are linearly independent within S. Let Λ_{11}^* be a $k \times k$ non-negative matrix whose elements are

$$\lambda_{ij} = \int_S \frac{f_i(x)f_j(x)}{f(x)} d\mu \qquad (i, j = 1, \ldots, k).$$

For any estimator $\hat{\theta}(x)$ we define a vector $\mathbf{d} = (d_1, \ldots, d_p)'$ by

$$d_i = \int_{S^c} \hat{\theta}(x)f_i(x)d\mu \qquad (i = 1, \ldots, p).$$

Suppose that $\mathbf{d}$ is given, then the unbiasedness condition is reduced to

$$\int_S \hat{\theta}(x)f_i(x)d\mu = c_i - d_i \qquad (i = 1, \ldots, p)$$

and under this condition we have from Theorem 2.1.1 that

$$\inf_{\substack{\hat{\theta} : \text{unbiased with} \\ \text{a fixed } \mathbf{d}}} \int_S \{\hat{\theta}(x)\}^2 f(x)d\mu = (\mathbf{c}_1 - \mathbf{d}_1)'\Lambda_{11}^{*-1}(\mathbf{c}_1 - \mathbf{d}_1),$$

where $\mathbf{c}' = (c_1, \ldots, c_p) = (\mathbf{c}_1', \mathbf{c}_2')$ and $\mathbf{d}' = (\mathbf{d}_1', \mathbf{d}_2')$ with $\mathbf{c}_1, \mathbf{d}_1 \in \mathbf{R}^k$ and $\mathbf{c}_2, \mathbf{d}_2 \in \mathbf{R}^{p-k}$. It is obvious that the set of possible vectors $\mathbf{d}_1$ is a linear subspace of $\mathbf{R}^k$ which is denoted by $\mathcal{D}_1$. Then we have

$$\inf_{\hat{\theta} : \text{unbiased}} \int_S \left\{\hat{\theta}(x)\right\}^2 f(x)d\mu = \min_{\mathbf{d}_1 \in \mathcal{D}_1}(\mathbf{c}_1 - \mathbf{d}_1)'\Lambda_{11}^{*-1}(\mathbf{c}_1 - \mathbf{d}_1).$$

If the dimension of $\mathcal{D}_1$ is ℓ, we have a basis $\mathbf{d}_1^* \ldots, \mathbf{d}_\ell^*$ of $\mathcal{D}_1$ and any $\mathbf{d}_1 \in \mathcal{D}_1$ can be expressed as

$$\mathbf{d}_1 = b_1\mathbf{d}_1^* + \cdots + b_\ell\mathbf{d}_\ell^* = D^*\mathbf{b},$$

where $D^* = (\mathbf{d}_1^*, \ldots, \mathbf{d}_\ell^*)$ is a $k \times \ell$ matrix and $\mathbf{b} = (b_1, \ldots, b_\ell)'$. Then it is straightforward to show that

$$\min_{\mathbf{d} \in \mathcal{D}_1}(\mathbf{c}_1 - \mathbf{d}_1)'\Lambda_{11}^{*-1}(\mathbf{c}_1 - \mathbf{d}_1)$$

$$= \min_{\mathbf{b}}(\mathbf{c}_1 - D^*\mathbf{b})'\Lambda_{11}^{*-1}(\mathbf{c}_1 - D^*\mathbf{b})$$

$$= \mathbf{c}_1'\Lambda_{11}^{*-1}\mathbf{c}_1 - \mathbf{c}_1'\Lambda_{11}^{*-1}D^*\left(D^{*'}\Lambda_{11}^{*-1}D^*\right)^{-1}D^{*'}\Lambda_{11}^{*-1}\mathbf{c}_1$$

and the minimum is attained by

$$\mathbf{b}^* = \left(D^{*'}\Lambda_{11}^{*-1}D^*\right)^{-1}D^{*'}\Lambda_{11}^{*-1}\mathbf{c}_1.$$

If, moreover, f_i $(i = 1, \ldots, p)$ are not linearly independent in S, the vector $\mathbf{c} - \mathbf{d}$ is restricted to a linear space E and

$$\inf_{\substack{\hat{\theta} : \text{unbiased with } \mathbf{d} \ ; \\ \mathbf{c} - \mathbf{d} \in E}} \int_S \left\{\hat{\theta}(x)\right\}^2 f(x)d\mu = (\mathbf{c}_0 - \mathbf{d}_0)'\Lambda_0^{*-1}(\mathbf{c}_0 - \mathbf{d}_0),$$

where Λ_0^* is the maximum rank principal minor of Λ_{11}^* and $\mathbf{c}_0$ and $\mathbf{d}_0$ are corresponding subvectors of $\mathbf{c}$ and $\mathbf{d}$. Then $\mathbf{d}_0$ is on a hyperplane H of some dimensionality and

$$\inf_{\hat{\theta}\,:\,\text{unbiased}} \int_S \left\{ \hat{\theta}(x) \right\}^2 f(x) d\mu = \min_{\mathbf{d}_0 \in H} (\mathbf{c}_0 - \mathbf{d}_0)' \Lambda_0^{*-1} (\mathbf{c}_0 - \mathbf{d}_0).$$

Note that even when the parameter space Θ is infinite, minimizing the variance with a finite number of conditions is often applied to obtain a lower bound for the variance of unbiased estimators, as is the case of Cramér-Rao, Bhattacharyya (See Isii (1964) with relation to mathematical programming).

2.2. BHATTACHARYYA TYPE BOUND FOR THE VARIANCE OF UNBIASED ESTIMATORS IN NON-REGULAR CASES

It is well known that the Cramér-Rao and Bhattacharyya bounds are most important and very useful for evaluating the variances of unbiased estimators (see, e.g. Rao (1973) and Lehmann (1983)). They are, however, not applicable to the non-regular cases when the support of the distribution is dependent on the parameter. Same is true about more general and simpler bounds by Hammersley (1950), Chapman and Robbins (1951), Kiefer (1952), Fraser and Guttman (1952), Fend (1959), Sen and Ghosh (1976) and Chatterji (1982), among others (For an exposition of some of this work along with extensions in different directions, see Polfeldt (1970a,b) and the papers of Víncze (1979), Khatri (1980) and Móri (1983), among others). In his paper, Polfeldt (1970a) discussed the lower bound for the variance of the unbiased estimators when the class of probability measures is one-sided, that is, when P_{θ_1} is absolutely continuous with respect to P_{θ_2} (symbolically, $P_{\theta_1} \ll P_{\theta_2}$) when $\theta_1 < \theta_2$ or $\theta_1 > \theta_2$. In this section, our main interest is to obtain the Bhattacharyya type bound when for any θ_1, θ_2 with $\theta_1 \neq \theta_2$, neither $P_{\theta_1} \ll P_{\theta_2}$ nor $P_{\theta_2} \ll P_{\theta_1}$.

Let $\mathcal{X}$ be an abstract sample space with x as its generic point, $\mathcal{A}$ a σ-field of subsets of $\mathcal{X}$, and let Θ be a parameter space assumed to be an open set in the real line. Let $\mathcal{P} = \{P_\theta | \theta \in \Theta\}$ be a class of probability measures on $(\mathcal{X}, \mathcal{A})$. We assume that for each $\theta \in \Theta$, $P_\theta(\cdot)$ is absolutely continuous with respect to a σ-finite measure μ. We denote $dP_\theta/d\mu$ by $f(x, \theta)$. For each $\theta \in \Theta$, we denote by $A(\theta)$ the set of points in $\mathcal{X}$ for which $f(x, \theta) > 0$.

We shall consider the Bhattacharyya type bound for the variance of unbiased estimators at some specified point θ_0 in Θ. We make the following assumptions :

(A.2.2.1)
$$\mu\left(\left(\bigcap_{c>0} \left(\bigcup_{|h|<c} A(\theta_0 + h) \right) \right) \Delta A(\theta_0) \right) = 0,$$

where $E \Delta F$ denotes the symmetric difference of two sets E and F.

(A.2.2.2) For every $\theta_0 \in \Theta$ there exist a positive number ε and a positive-valued function $\rho(x)$ such that for every $\theta \in (\theta_0 - \varepsilon, \theta_0 + \varepsilon)$, $\rho(x) > f(x, \theta)$ for all $x \in A(\theta)$, and for every $\theta \in (\theta_0 - \varepsilon, \theta_0 + \varepsilon)$

$$\int_{A(\theta)} |\gamma(x)| f(x,\theta) d\mu < \infty \quad \text{implies} \quad \int_{\substack{\cup A(\theta) \\ \theta \in (\theta_0 - \varepsilon, \theta_0 + \varepsilon)}} |\gamma(x)| \rho(x) d\mu < \infty.$$

(A.2.2.3) For some positive integer k

$$\overline{\lim_{h \to 0}} \sup_{x \in \cup_{j=1}^i A(\theta_0 + jh) - A(\theta_0)} \frac{\left| \sum_{j=1}^i (-1)^j \binom{i}{j} f(x, \theta_0 + jh) \right|}{|h|^j \rho(x)} < \infty, \quad i = 1, \ldots, k.$$

First we prove the following lemma.

Lemma 2.2.1. *Assume that the conditions (A.2.2.1) and (A.2.2.3) hold. If $\varphi(x)$ is any measurable function for which $\int_\mathcal{X} |\varphi(x)| \rho(x) d\mu < \infty$, then*

$$\lim_{h \to 0} \frac{1}{h^i} \int_{\cup_{j=1}^i A(\theta_0 + jh) - A(\theta_0)} \sum_{j=1}^i (-1)^j \binom{i}{j} \varphi(x) f(x, \theta_0 + jh) d\mu = 0.$$

Proof. By (A.2.2.3) it follows that for every $i = 1, \ldots, k$ and every $\theta_0 \in \Theta$, there exist positive number ε and k_i such that

$$\frac{1}{|h|^i} \left| \sum_{j=1}^i (-1)^j \binom{i}{j} f(x, \theta_0 + jh) \right| < k_i \rho(x) \quad \text{for} \quad |ih| < \varepsilon \quad \text{and} \quad x \in A(\theta_0)^c,$$

where $A(\theta_0)^c$ denotes the complement of the set $A(\theta_0)$. Also, it follows from (A.2.2.1) that for every $j = 1, \ldots, i$, there exist a sequence $\{\varepsilon_{jn}\}$ of positive numbers converging to zero as $n \to \infty$ and a monotone non-increasing sequence $\{S_{jn}\}$ of measurable sets such that $|jh| < \varepsilon_{jn}$ implies $A(\theta_0 + jh) - A(\theta_0) \subset S_{jn}$ and $\mu\left(\bigcap_{n=1}^\infty S_{jn}\right) = 0$. If for $j = 1, \ldots, i$, $|jh| < \varepsilon_{jn}$, then

$$\left| \frac{1}{h^i} \int_{\cup_{j=1}^i A(\theta_0 + jh) - A(\theta_0)} \sum_{j=1}^i (-1)^j \binom{i}{j} \varphi(x) f(x, \theta_0 + jh) d\mu \right|$$

$$\leq \int_{\cup_{j=1}^i A(\theta_0 + jh) - A(\theta_0)} k_i |\varphi(x)| \rho(x) d\mu \leq \int_{\cup_{j=1}^i S_{jn}} k_i |\varphi(x)| \rho(x) d\mu$$

$$\leq \sum_{j=1}^i \int_{S_{jn}} k_i |\varphi(x)| \rho(x) d\mu,$$

which tends to zero as $n \to \infty$. The proof follows.

Remark 2.2.1. The assumption (A.2.2.2) together with the condition in Lemma 2.2.1 is satisfied with $p(x) = \sum_{i=1}^\infty c_i f(x, \theta_i)$ when the following holds : For each θ_0 and $\varepsilon > 0$, there exist countable points $\theta_1, \theta_2, \ldots$ and positive constants $c_1, c_2, \ldots$ such that $\cup_{i=1}^\infty A(\theta_i) \supset A(\theta)$ for all $\theta \in (\theta_0 - \varepsilon, \theta_0 + \varepsilon)$ and that $\sum_{i=1}^\infty c_i < \infty$ and $\sum_{i=1}^\infty c_i f(x, \theta_i) > f(x, \theta)$ for all $\theta \in (\theta_0 - \varepsilon, \theta_0 + \varepsilon)$ and almost all $x [\mu]$.

We assume the following.

(A.2.2.4) For each $x \in A(\theta_0)$, $f(x, \theta)$ is k-times continuously differentiable in θ at $\theta = \theta_0$.

(A.2.2.5) For each $i = 1, \ldots, k$,

$$\varlimsup_{\substack{h \to 0 \\ x \in A(\theta_0)}} \sup \frac{\left| \frac{\partial^i}{\partial \theta^i} f(x, \theta_0 + h) \right|}{\rho(x)} < \infty, \quad \text{where } \rho(x) \text{ is defined in (A.2.2.2).}$$

We now prove the following theorem on the Bhattacharyya type bound for the variance of unbiased estimators.

Theorem 2.2.1. *Assume that the conditions (A.2.2.1) to (A.2.2.5) hold. Let $g(\theta)$ be an estimable function which is k-times differentiable over Θ. Let $\hat{g}(x)$ be an unbiased estimator of $g(\theta)$ satisfying*

$$\int_{A(\theta_0)} |\hat{g}(x)| \rho(x) d\mu < \infty.$$

Further, let Λ be a $k \times k$ non-negative definite matrix whose elements are

$$\lambda_{ij} = \int_{A(\theta_0)} \frac{1}{f(x, \theta_0)} \left\{ \frac{\partial^i f(x, \theta_0)}{\partial \theta^i} \frac{\partial^j f(x, \theta_0)}{\partial \theta^j} \right\} d\mu, \quad i, j = 1, \ldots, k.$$

Assume that $\lambda_{ii}, i = 1, \ldots, k$ are finite. If Λ is non-singular at θ_0, then

$$(2.2.1) \qquad V_{\theta_0}(\hat{g}) \geq \left(g^{(1)}(\theta_0), \ldots, g^{(k)}(\theta_0) \right) \Lambda^{-1} \left(g^{(1)}(\theta_0), \ldots, g^{(k)}(\theta_0) \right)',$$

where $g^{(i)}(\theta)$ is the i-th order derivative of $g(\theta)$.

Proof. Denote

$$g_0(\theta) = \int_{A(\theta_0)} \hat{g}(x) f(x, \theta) d\mu \quad \text{and} \quad \Delta_h f(x, \theta) = f(x, \theta + h) - f(x, \theta).$$

Then, by (A.2.2.2) and (A.2.2.5), we have

$$(2.2.2) \qquad \left[\frac{\partial^i}{\partial \theta^i} g_0(\theta) \right]_{\theta = \theta_0} = \lim_{h \to 0} \frac{1}{h^i} \Delta_h^i g_0(\theta_0) = \lim_{h \to 0} \frac{1}{h^i} \int_{A(\theta_0)} \hat{g}(x) \Delta_h^i f(x, \theta_0) d\mu$$

$$= \lim_{h \to 0} \int_{A(\theta_0)} \hat{g}(x) \frac{1}{h^i} \Delta_h^i f(x, \theta_0) d\mu$$

$$= \lim_{h \to 0} \int_{A(\theta_0)} \hat{g}(x) \left\{ \frac{\partial^i}{\partial \theta^i} f(x, \theta_0 + \xi h) \right\} d\mu$$

$$= \int_{A(\theta_0)} \hat{g}(x) \left[\frac{\partial^i}{\partial \theta^i} f(x, \theta) \right]_{\theta = \theta_0} d\mu,$$

where $\Delta_h^i g(\theta) = \Delta_h^{i-1}(\Delta_h g(\theta)), i = 1, \ldots, k$ and $0 < \xi < 1$. Since $g(\theta) = \int_{A(\theta)} \hat{g}(x) f(x, \theta) d\mu$, we obtain for each $i = 1, \ldots, k$

$$(2.2.3)\Delta_h^i \left(g(\theta_0) - g_0(\theta_0)\right) = \sum_{j=1}^{i} (-1)^j \binom{i}{j} \{g(\theta_0 + jh) - g_0(\theta_0 + jh)\}$$

$$= \sum_{j=1}^{i} (-1)^j \binom{i}{j} \left\{ \int_{A(\theta_0+jh)} \hat{g}(x) f(x, \theta_0 + jh) d\mu - \int_{A(\theta_0)} \hat{g}(x) f(x, \theta_0 + jh) d\mu \right\}$$

$$= \sum_{j=1}^{i} (-1)^j \binom{i}{j} \left\{ \left(\int_{A(\theta_0+jh)} - \int_{A(\theta_0) \cap A(\theta_0+jh)} - \int_{A(\theta_0) - A(\theta_0+jh)} \right) \right.$$

$$\left. \hat{g}(x) f(x, \theta_0 + jh) d\mu \right\}$$

$$= \sum_{j=1}^{i} (-1)^j \binom{i}{j} \left\{ \left(\int_{A(\theta_0+jh)} - \int_{A(\theta_0) \cap A(\theta_0+jh)} \right) \hat{g}(x) f(x, \theta_0 + jh) d\mu \right\}$$

$$= \sum_{j=1}^{i} (-1)^j \binom{i}{j} \int_{A(\theta_0+jh) - A(\theta_0)} \hat{g}(x) f(x, \theta_0 + jh) d\mu$$

$$= \sum_{j=1}^{i} (-1)^j \binom{i}{j} \int_{\cup_{k=1}^{i} A(\theta_0+kh) - A(\theta_0)} \hat{g}(x) f(x, \theta_0 + jh) d\mu$$

$$= \int_{\cup_{k=1}^{i} A(\theta_0+kh) - A(\theta_0)} \sum_{j=1}^{i} (-1)^j \binom{i}{j} \hat{g}(x) f(x, \theta_0 + jh) d\mu.$$

By (2.2.3) and Lemma 2.2.1, we have for each $i = 1, \ldots, k$,

$$(2.2.4) \left[\frac{\partial^i}{\partial \theta^i} g(\theta) \right]_{\theta=\theta_0} = \lim_{h \to 0} \frac{1}{h^i} \Delta_h^i g(\theta_0) = \lim_{h \to 0} \frac{1}{h^i} \Delta_h^i \left(g(\theta_0) - g_0(\theta_0)\right) + \lim_{h \to 0} \frac{1}{h^i} \Delta_h^i g_0(\theta_0)$$

$$= \lim_{h \to 0} \frac{1}{h^i} \Delta_h^i g_0(\theta_0) = \left[\frac{\partial^i}{\partial \theta^i} g_0(\theta) \right]_{\theta=\theta_0}.$$

From (2.2.2) and (2.2.4) we obtain for each $i = 1, \ldots, k$,

$$(2.2.5) \qquad \left[\frac{\partial^i}{\partial \theta^i} g(\theta) \right]_{\theta=\theta_0} = \int_{A(\theta_0)} \hat{g}(x) \left[\frac{\partial^i}{\partial \theta^i} f(x, \theta) \right]_{\theta=\theta_0} d\mu.$$

Proceeding now as in the regular case (see, e.g. Zacks (1971, page 190)), one can show that the Bhattacharyya bound for the variance of unbiased estimators of $g(\theta)$ is given by (2.2.1).

Example 2.2.1. We consider the location parameter case, i.e., $f(x, \theta) = f_0(x - \theta)$, and unbiased estimators of θ. We assume that for any $p \geq 1$, the density function $f_0(x)$ is given by

$$f_0(x) = \begin{cases} c(1 - x^2)^{p-1} & \text{if } |x| < 1, \\ 0 & \text{if } |x| \geq 1, \end{cases}$$

where $c = 1/B(1/2, p)$ where $B(\alpha, \beta) = \int_0^1 x^{\alpha-1}(1-x)^{\beta-1}dx \quad (\alpha > 0, \beta > 0)$.

Case (i). Let $p = 1$. Then the distribution is uniform, and it is easy to check that $\min_{\hat{\theta}:\text{unbiased}} V_{\theta_0}(\hat{\theta}) = 0$ for any specific value θ_0(see also Takeuchi, 1961b).

Case (ii). Let $p = 2$. In this case, it is easy to check that the Fisher information $\int_{-1}^1 \{f_0'(x)/f_0(x)\}^2 f_0(x)dx = \infty$. For any $\varepsilon > 0$, we define an estimator $\hat{\theta}_\varepsilon$ which satisfies

$$\hat{\theta}_\varepsilon(x) = \begin{cases} -c_\varepsilon f_0'(x)/f_0(x) & \text{if } |x| \leq 1 - \varepsilon, \\ 0 & \text{if } 1 - \varepsilon < |x| \leq 1, \end{cases}$$

where c_ε is a constant determined from the equations

(2.2.6) $$\int_{-1}^1 \hat{\theta}_\varepsilon(x)f_0(x)dx = 0 \quad \text{and} \quad \int_{-1}^1 \hat{\theta}_\varepsilon(x)f_0'(x)dx = -1.$$

We shall determine $\hat{\theta}_\varepsilon(x)$ for x outside the interval $[-1, 1]$ from the unbiasedness condition

(2.2.7) $$\int_{-1+\theta}^{1+\theta} \hat{\theta}_\varepsilon(x)f_0(x - \theta)dx = \theta.$$

First consider the case $0 < \theta \leq 1$, and define

(2.2.8) $$g(\theta) = \int_{-1+\theta}^1 \hat{\theta}_\varepsilon(x)f_0(x - \theta)dx.$$

Since $\hat{\theta}_\varepsilon(x)$ and $f_0'(x)$ are bounded, $g(\theta)$ is differentiable and $g'(\theta) = -\int_{-1+\theta}^1 \hat{\theta}_\varepsilon(x)f_0'(x-\theta)dx$.

If we assume that $\hat{\theta}_\varepsilon(x)$ is bounded for $1 < x \leq 2$, the right hand side of (2.2.8) is also differentiable, and we have by (2.2.7) and (2.2.8) that

(2.2.9) $$1 - g'(\theta) = -\int_1^{1+\theta} \hat{\theta}_\varepsilon(x)f_0'(x - \theta)dx.$$

Differentiating (2.2.9) again, and noting that $\lim_{x \to 1-0} f_0'(x) = -3/2$, we have

(2.2.10) $$g''(\theta) = -\frac{3}{2}\hat{\theta}_\varepsilon(1 + \theta) - \int_1^{1+\theta} \hat{\theta}_\varepsilon(x)f_0''(x - \theta)dx.$$

If $\hat{\theta}_\varepsilon$ satisfies (2.2.10), then it also satisfies (2.2.9) since $\lim_{\theta \to 0} g'(\theta) = -1$; it also satisfies (2.2.8) since $\lim_{\theta \to 0} g(\theta) = 0$ by (2.2.6).

Since the integral equation (2.2.10) is of Volterra's second type, it follows that the solution $\hat{\theta}_\varepsilon(x)$ exists and is bounded. Repeating the same process, we have the solution $\hat{\theta}_\varepsilon(x)$ for all $x > 1$. Similarly, we can construct $\hat{\theta}_\varepsilon(x)$ for $x < -1$. On the other hand,

$$(2.2.11) \qquad V_{\theta_0}(\hat{\theta}_\varepsilon) = c_\varepsilon^2 \int_{-1+\varepsilon}^{1-\varepsilon} \{f_0'(x)/f_0(x)\}^2 f_0(x)dx,$$

while, from (2.2.6), we have

$$(2.2.12) \qquad c_\varepsilon \int_{-1+\varepsilon}^{1-\varepsilon} f_0'(x)dx = 0 \quad \text{and} \quad c_\varepsilon \int_{-1+\varepsilon}^{1-\varepsilon} \{f_0'(x)/f_0(x)\}^2 f_0(x)dx = 1.$$

Hence $V_{\theta_0}(\hat{\theta}_\varepsilon) = c_\varepsilon$. Now, since

$$\lim_{\varepsilon \to 0} \int_{-1+\varepsilon}^{1-\varepsilon} \{f_0'(x)/f_0(x)\}^2 f_0(x)dx = \infty,$$

we have from (2.2.12) $\lim_{\varepsilon \to 0} c_\varepsilon = 0$. Consequently

$$\inf_{\hat{\theta}\,:\,\text{unbiased}} V_{\theta_0}(\hat{\theta}_\varepsilon) = 0$$

for any specified value θ_0.

Before proceeding on to the next cases, we note the following : If $k < p/2$, then λ_{ii} ($i = 1, \ldots, k$) given in Theorem 2.2.1 are finite, since $\int_{-1}^{1}(1 - x^2)^{p-2k-1}dx < \infty$. Also,

$$(2.2.13) \qquad \lambda_{ij} = \int_{\theta-1}^{\theta+1} \frac{1}{f_0(x-\theta)} \frac{\partial^i f_0(x-\theta)}{\partial \theta^i} \frac{\partial^j f_0(x-\theta)}{\partial \theta^j} dx$$

$$= \int_{-1}^{1} (-1)^{i+j} \frac{1}{f_0(x)} f_0^{(i)}(x) f_0^{(j)}(x)dx \quad ; \quad i, j = 1, \ldots, k.$$

We also obtain for $|x| < 1$,

$$(2.2.14) \qquad \begin{aligned} f_0^{(1)}(x) &= -2c(p-1)x(1-x^2)^{p-2}; \\ f_0^{(2)}(x) &= -2c(p-1)\left\{(1-x^2)^{p-2} - 2(p-2)x^2(1-x^2)^{p-3}\right\}; \\ f_0^{(3)}(x) &= 4c(p-1)(p-2)\left\{3x(1-x^2)^{p-3} - 2(p-3)x^3(1-x^2)^{p-4}\right\}; \end{aligned}$$

If $i + j$ is an odd number, it follows by (2.2.13) and (2.2.14) that $\lambda_{ij} = 0$ since

$f_0^{(i)}(x)f_0^{(j)}(x)$ is an odd function. From (2.2.13) and (2.2.14), we have

$$\lambda_{11} = 4c(p-1)^2 B\left(3/2, p-2\right);$$

$$\lambda_{13} = 8c(p-1)^2(p-2)\left\{2(p-3)B\left(5/2, p-4\right) - 3B\left(3/2, p-3\right)\right\};$$

$$\lambda_{22} = 4c(p-1)^2\left\{B\left(1/2, p-2\right) - 4(p-2)B\left(3/2, p-3\right) + 4(p-2)^2 B\left(5/2, p-4\right)\right\};$$

$$\lambda_{33} = 16c(p-1)^2(p-2)^2\left\{9B\left(3/2, p-4\right) - 12(p-3)B\left(5/2, p-5\right)\right.$$

$$\left. + 4(p-3)^2 B\left(7/2, p-6\right)\right\};$$

...

Case (iii). Let $p = 3, 4$. Then, we have, for any unbiased estimator $\hat{\theta}(X)$ of θ,

(2.2.15)
$$\int_{-1}^{1} \hat{\theta}(x)f_0(x)dx = 0,$$

(2.2.16)
$$\int_{-1}^{1} \hat{\theta}(x)f_0'(x)dx = 1,$$

(2.2.17)
$$\int_{-1}^{1} \hat{\theta}(x)f_0^{(k)}(x)dx = 0 \quad (k = 2,\ldots,p-1).$$

Noting that

$$\int_{-1}^{1}\left[\left\{\sum_{k=1}^{p-1} c_k f_0^{(k)}(x)\right\}^2 \bigg/ f_0(x)\right]dx < \infty$$

implies $c_2 = \cdots = c_{p-1} = 0$, we have by Theorem 2.1.1 that the infimum of

$$V_0(\hat{\theta}) = \int_{-1}^{1}\hat{\theta}^2(x)f_0(x)dx$$

under (2.2.15), (2.2.16) and (2.2.17) is given by

$$\inf_{\hat{\theta}\,:\,(2.2.15)\sim(2.2.17)} V_0(\hat{\theta}) = 1/\lambda_{11},$$

where

$$\lambda_{11} = \int_{-\infty}^{\infty}\left\{f_0'(x)/f_0(x)\right\}^2 f_0(x)dx = (p-1)(2p-1)/(p-2)$$

and for any $\varepsilon > 0$ there exists $\hat{\theta}_\varepsilon(x)$ in $(-1,1)$ and satisfying (2.2.15), (2.2.16) and (2.2.17), and

$$\int_{-1}^{1}\hat{\theta}_\varepsilon^2(x)f_0(x)dx < \frac{1}{\lambda_{11}} + \varepsilon.$$

We can extend $\hat{\theta}_\varepsilon(x)$ for x outside $(-1, 1)$ from the unbiasedness condition

$$\int_{-1+\theta}^{1+\theta} \hat{\theta}(x) f_0(x - \theta) dx = \theta.$$

First, we consider the case when $0 \le \theta < 1$, and define $g(\theta) = \int_{-1+\theta}^{1} \hat{\theta}_\varepsilon(x) f_0(x - \theta) dx$. In a similar way to the case (ii), we have for $k = 2, \ldots, p - 1$,

$$(2.2.18) \qquad (-1)^{k+1} g^{(k+1)}(\theta) = B_k \hat{\theta}_\varepsilon(1 + \theta) - \int_{1}^{1+\theta} \hat{\theta}_\varepsilon(x) f_0^{(k+1)}(x - \theta) dx,$$

where $B_k = \lim_{x \to 1-0} f_0^{(k)}(x)$. Since the integral equation (2.2.18) is again of Volterra's second type, it follows that the solution $\hat{\theta}_\varepsilon(x)$ exists. Repeating the process, we can construct an unbiased estimator $\hat{\theta}_\varepsilon(x)$ for all values of x. Then, it follows that

$$\inf_{\hat{\theta} : \text{unbiased}} V_{\theta_0}(\hat{\theta}) = 1/\lambda_{11}$$

for any specified value θ_0.

Case (iv). Let $p = 5, 6$. Noting that

$$\int_{-1}^{1} \{f_0'(x) f_0''(x)/f_0(x)\} \, dx = 0,$$

$$\int_{-1}^{1} \{f_0''(x)/f_0(x)\}^2 f_0(x) dx = \frac{(p-1)(2p-1)(2p-3)}{(p-2)(p-3)(p-4)}(2p^2 - 7p + 8) = \lambda_{22} \quad \text{(say)},$$

and

$$\int_{-1}^{1} \left\{ \sum_{k=1}^{p-1} c_k f_0^{(k)}(x)/f_0(x) \right\}^2 f_0(x) dx < \infty,$$

imply $c_3 = \cdots = c_{p-1} = 0$, we see that

$$V_{\theta_0}(\hat{\theta}) \ge (1, 0) \begin{pmatrix} \lambda_{11} & 0 \\ 0 & \lambda_{22} \end{pmatrix}^{-1} \begin{pmatrix} 1 \\ 0 \end{pmatrix} = 1/\lambda_{11}$$

for any specified θ_0, where λ_{11} is defined above (2.2.13). Here again, as in the previous case,

$$\inf_{\hat{\theta} : \text{unbiased}} V_{\theta_0}(\hat{\theta}) = 1/\lambda_{11}$$

for any specific θ_0.

Case (v). Let $p = 7$. In this case, we see that $k = 3$. Using Theorem 2.2.1, (2.2.13) and (2.2.14), we obtain

$$V_{\theta_0}(\hat{\theta}) \ge \frac{1}{|\Lambda|} \begin{vmatrix} \lambda_{22} & 0 \\ 0 & \lambda_{33} \end{vmatrix} = \left\{ \lambda_{11} \left(1 - \frac{\lambda_{13}^2}{\lambda_{11}\lambda_{33}} \right) \right\}^{-1}$$

where

$$\Lambda = \begin{pmatrix} \lambda_{11} & 0 & \lambda_{13} \\ 0 & \lambda_{22} & 0 \\ \lambda_{13} & 0 & \lambda_{33} \end{pmatrix}$$

with

$$\lambda_{11} = 144cB\,(3/2,5)\,;$$
$$\lambda_{13} = 1440c\,\{8B\,(5/2,3) - 3B\,(3/2,4)\}\,;$$
$$\lambda_{22} = 144c\,\{B\,(1/2,5) - 20B\,(3/2,4) + 100B\,(5/2,3)\}\,;$$
$$\lambda_{33} = 14400c\,\{9B\,(3/2,5) - 48B\,(5/2,2) + 64B\,(7/2,1)\}\,.$$

We also obtain

$$\frac{\lambda_{13}^2}{\lambda_{11}\lambda_{33}} = \frac{\{8B\,(5/2,3) - 3B\,(3/2,4)\}^2}{B\,(3/2,5)\,\{9B\,(3/2,3) - 48B\,(5/2,2) + 64B\,(7/2,1)\}}.$$

Here again

$$\inf_{\hat{\theta}\,:\,\text{unbiased}} V_{\theta_0}(\hat{\theta}) = \left\{\lambda_{11}\left(1 - \frac{\lambda_{13}^2}{\lambda_{11}\lambda_{33}}\right)\right\}^{-1}$$

for any specific θ_0, i.e. we have a sharp bound.

Case (vi). For $p \geq 8$ we can continue in a similar manner by choosing $k = [(p-1)/2]$, where $[s]$ denotes the largest integer less than or equal to s.

The above discussion establishes that here the bound is sharp but generally it is not attainable.

2.3. LOWER BOUND FOR THE VARIANCE OF UNBIASED ESTIMATORS FOR ONE-DIRECTIONAL DISTRIBUTIONS

For the lower bound of the variance of unbiased estimators, the most famous is the so-called Cramér-Rao bound. But the Cramér-Rao bound and its Bhattacharyya extension assume a set of regularity conditions. Chapman and Robbins (1951), Kiefer (1952) and Fraser and Guttman (1952) obtained bounds with much less stringent assumptions, but they still require the independence of the support of the parameter θ or almost equivalently that the distribution with $\theta \neq \theta_0$ is absolutely continuous with respect to that with $\theta = \theta_0$ when θ_0 is the specified parameter value at which the variance is evaluated. In the non-regular cases the Cramér-Rao bound has been discussed by Víncze (1979), Móri (1983) and others.

In Section 2.2 we get the Bhattacharyya type bound for the variance of unbiased estimators in the non-regular cases. In this section we introduce the concept of the one-directionality which includes both the case of location (and scale) parameter and selection parameter and also other cases, and show that the bound for the variance of unbiased estimators is sharp in the sense that the actual infimum of the variance of unbiased estimators is equal to the bound for a specified θ_0 for this class of non-regular distributions. We also establish that for a wide class of the non-regular distributions the infimum of the variance of unbiased estimators can be zero when the sample size is not smaller than 2.

A simple but rather general case of the class of distributions of which none dominates another is that it is characterized by a real parameter θ and the distribution shifts monotonically as θ changes. For one dimensional random variables the case can be visualized as follows.

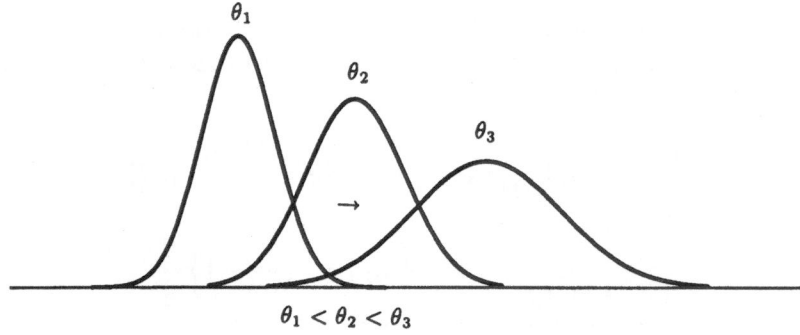

$$\theta_1 < \theta_2 < \theta_3$$

Figure 2.3.1. An example of one-directional distributions

Next mathematical definition for such cases in a rather general set-up is given and is termed as one-directional family of distributions.

Definition of the one-directional distribution

We assume that we are given a model consisting of a sample space $(\mathcal{X}, \mathcal{B})$ and a family $\mathcal{P} = \{P_\theta : \theta \in \Theta\}$ of probability measures, where a parameter space Θ is an open subset in a Euclidean 1-space $\mathbf{R}^1$. We make the following conditions (B1)–(B5).

(B.1) For each $\theta \in \Theta$ P_θ is absolutely continuous with respect to a σ-finite measure μ and the corresponding density w.r.t. μ is $f(x, \theta)$.

Let $A(\theta)$ be a support of $f(x, \theta)$, that is, $A(\theta) = \{x | f(x, \theta) > 0\}$. The determination of $A(\theta)$ is not unique in so far as any null set may be added to it, but in the sequel we take one and fixed determination of $A(\theta)$ for every $\theta \in \Theta$ which satisfies the following :

(B.2) For any disjoint points θ_1 and θ_2 in Θ, neither $A(\theta_1) \supset A(\theta_2)$ nor $A(\theta_1) \subset A(\theta_2)$.

(B.3) For $\theta_1 < \theta_2 < \theta_3$,

$$A(\theta_1) \cap A(\theta_3) \subset A(\theta_1) \cap A(\theta_2) , \quad A(\theta_1) \cap A(\theta_3) \subset A(\theta_2) \cap A(\theta_3).$$

(B.4) If θ_n tends to θ as $n \to \infty$, then

$$\mu\left(\left(\bigcup_{n=1}^{\infty} \bigcap_{i \geq n} A(\theta_i)\right) \triangle A(\theta)\right) = \mu\left(\left(\bigcap_{n=1}^{\infty} \bigcup_{i \geq n} A(\theta_i)\right) \triangle A(\theta)\right) = 0,$$

where $E \triangle F$ denotes the symmetric difference of two sets E and F.

(B.5) For any two points θ_1 and θ_2 in Θ with $\theta_1 < \theta_2$, there exists a finite number of ξ_i $(i = 0, 1, \ldots, k)$ such that $\theta_1 = \xi_0 < \xi_1 < \cdots < \xi_k = \theta_2$ and $\mu(A(\xi_i) \cap A(\xi_{i-1})) > 0$ $(i = 1, \ldots, k)$.

Then $\mathcal{P}$ is called to be a one-directional family of distributions if the conditions (B.1) to (B.5) hold.

The lower bound for the variance of unbiased estimators when X is a real random variable

Now suppose that X is a real random variable with a density function $f(x, \theta)$ whose support is an open interval $(a(\theta), b(\theta))$, then the condition of one-directionality means that $a(\theta)$ and $b(\theta)$ are both monotone increasing (or decreasing) and continuous functions. Without loss of generality we assume that $a(\theta)$ and $b(\theta)$ are monotone increasing functions. Therefore we can formulate the following problem. Let $\mathcal{X} = \Theta = \mathbf{R}^1$. Let X be a real random variable with a density function $f(x, \theta)$ (with respect to the Lebesgue measure μ) satisfying the following conditions (A.2.3.1) to (A.2.3.7) :

(A.2.3.1)
$$f(x, \theta) > 0 \quad \text{for} \quad a(\theta) < x < b(\theta),$$
$$f(x, \theta) = 0 \quad \text{for} \quad x \le a(\theta), \ x \ge b(\theta),$$

where $f(x, \theta)$ is continuous in x and θ for which $a(\theta) < x < b(\theta)$, and $(p+1)$-times continuously differentiable in θ for almost all x $[\mu]$ for some non-negative integer p both of functions $a(\theta)$ and $b(\theta)$ are p-times continuously differentiable and $a'(\theta) > 0$, $b'(\theta) > 0$ for all θ.

(A.2.3.2)
$$\lim_{x \to a(\theta)+0} f(x, \theta) = \lim_{x \to b(\theta)-0} f(x, \theta) = 0,$$

and for some positive integer p

$$\lim_{x \to a(\theta)+0} \frac{\partial^i}{\partial \theta^i} f(x, \theta) = \lim_{x \to b(\theta)-0} \frac{\partial^i}{\partial \theta^i} f(x, \theta) = 0 \quad (i = 1, \ldots, p-1),$$

$$\lim_{x \to a(\theta)+0} \frac{\partial^p}{\partial \theta^p} f(x, \theta) = A_p(\theta), \quad \lim_{x \to b(\theta)-0} \frac{\partial^p}{\partial \theta^p} f(x, \theta) = B_p(\theta),$$

where $A_p(\theta)$ and $B_p(\theta)$ are non-zero, finite and continuous in θ.

(A.2.3.3) $(\partial^i/\partial \theta^i) f(x, \theta)$ $(i = 1, \ldots, p)$ are linearly independent.

(A.2.3.4) For $\theta_0 \in \Theta$

$$0 < \int_{a(\theta_0)}^{b(\theta_0)} \frac{\left\{ \frac{\partial^i}{\partial \theta^i} f(x, \theta_0) \right\}^2}{f(x, \theta_0)} d\mu(x)$$

is finite for each $i = 1, \ldots, k$ and

$$\int_{a(\theta_0)}^{b(\theta_0)} \frac{\left\{ \sum_{i=k+1}^{p} c_i \frac{\partial^i}{\partial \theta^i} f(x, \theta_0) \right\}^2}{f(x, \theta_0)} d\mu(x)$$

is infinite for each $i = k+1, \ldots, p$ unless $c_{k+1} = \cdots = c_p = 0$.

(A.2.3.5) For $\theta_0 \in \Theta$ there exist a positive number ε and a positive-valued measurable function $\rho(x)$ such that for every $x \in A(\theta)$ and every $\theta \in (\theta_0 - \varepsilon, \theta_0 + \varepsilon)$, $\rho(x) > f(x, \theta)$, and for every $\theta \in (\theta_0 - \varepsilon, \theta_0 + \varepsilon)$, $\int_{A(\theta)} |\gamma(x)| f(x, \theta) d\mu < \infty$ implies $\int_{\underset{\theta \in (\theta_0 - \varepsilon, \theta_0 + \varepsilon)}{\cup} A(\theta)} |\gamma(x)| \rho(x) d\mu < \infty$.

(A.2.3.6) For each $i = 1, \ldots, p+1$,

$$\varlimsup_{h \to 0} \sup_{x \in \underset{j=1}{\overset{i}{\cup}} A(\theta_0 + jh) - A(\theta_0)} \frac{\left| \sum_{j=1}^{i} (-1)^j \binom{i}{j} f(x, \theta_0 + jh) \right|}{|h|^i \rho(x)} < \infty.$$

(A.2.3.7) For each $i = 1, \ldots, p+1$,

$$\varlimsup_{h \to 0} \sup_{x \in A(\theta_0)} \frac{\left| \frac{\partial^i}{\partial \theta^i} f(x, \theta_0 + h) \right|}{\rho(x)} < \infty.$$

Note that the conditions (A.2.3.5), (A.2.3.6) and (A.2.3.7) are assumed to obtain the Bhattacharyya bound for the variance of unbiased estimators (see Section 2.2).

First we consider the special case when $p = 0$, then we have to modify slightly the condition (A.2.3.2) as follows :

(A.2.3.2)* $\lim_{x \to a(\theta)+0} f(x, \theta) = A_0(\theta) > 0, \qquad \lim_{x \to b(\theta)-0} f(x, \theta) = B_0(\theta) > 0.$

In the following theorem we shall have a lower bound.

Theorem 2.3.1. *Let $g(\theta)$ be continuously differentiable over Θ. Let $\hat{g}(x)$ be an unbiased estimator of $g(\theta)$. If, for $p = 0$ and a fixed θ_0, the conditions (A.2.3.1), (A.2.3.2)*, (A.2.3.5), (A.2.3.6) and (A.2.3.7) hold, then*

$$\min_{\hat{g} : \text{unbiased}} V_{\theta_0}(\hat{g}) = 0,$$

where $V_{\theta_0}(\hat{g})$ denotes the variance of $\hat{g}$ at $\theta = \theta_0$.

Proof. We first define

$$\hat{g}(x) = g(\theta_0) \qquad \text{for} \quad a(\theta_0) < x < b(\theta_0).$$

Since $a(\theta_0) < a(\theta) < b(\theta_0) < b(\theta)$ for $\theta > \theta_0$, it follows from the unbiasedness condition that

$$(2.3.1) \qquad g(\theta) = \int_{a(\theta)}^{b(\theta)} \hat{g}(x) f(x, \theta) d\mu = \int_{a(\theta)}^{b(\theta_0)} \hat{g}(x) f(x, \theta) d\mu + \int_{b(\theta_0)}^{b(\theta)} \hat{g}(x) f(x, \theta) d\mu.$$

Putting

$$h(\theta) = \int_{a(\theta)}^{b(\theta_0)} \hat{g}(x) f(x, \theta) d\mu,$$

we have by (2.3.1)

(2.3.2) $$\int_{b(\theta_0)}^{b(\theta)} \hat{g}(x) f(x, \theta) d\mu = g(\theta) - h(\theta).$$

Differentiating both sides of (2.3.2) with respect to θ, we obtain

(2.3.3) $$b'(\theta) B_0(\theta) \hat{g}(b(\theta)) + \int_{b(\theta_0)}^{b(\theta)} \hat{g}(x) \left\{ \frac{\partial}{\partial \theta} f(x, \theta) \right\} d\mu = g'(\theta) - h'(\theta).$$

Differentiation under the integral sign is admitted because of (A.2.3.7) with $p = 0$. If $\hat{g}(x)$ satisfies (2.3.3), then it also satisfies (2.3.1) since $g(\theta_0) = h(\theta_0)$. Since, by (A.2.3.1) and (A.2.3.2)*, $b'(\theta) B_0(\theta) > 0$, it follows that the integral equation (2.3.3) is of Voterra's second type, hence the solution $\hat{g}(x)$ exists for $b(\theta) > x \geq b(\theta_0)$. Similarly we can construct $\hat{g}(x)$ for $x < a(\theta_0)$. Repeating the same process we can define $\hat{g}(x)$ for all x. Hence we have

$$\min_{\hat{g} \,:\, \text{unbiased}} V_{\theta_0}(\hat{g}) = 0$$

Thus we complete the proof.

The following useful lemma is a special case of the result of Section 2.1.

Lemma 2.3.1. *Let $g(\theta)$ be p-times differentiable over Θ. Suppose that the conditions (A.2.3.3) and (A.2.3.4) hold and G is the class of the all estimators $\hat{g}(X)$ of $g(\theta)$ for which*

$$\int_{a(\theta_0)}^{b(\theta_0)} \hat{g}(x) f(x, \theta_0) d\mu = g(\theta_0)$$

$$\int_{a(\theta_0)}^{b(\theta_0)} \hat{g}(x) \left\{ \frac{\partial^i}{\partial \theta^i} f(x, \theta_0) \right\} d\mu = g^{(i)}(\theta_0) \qquad (i = 1, \ldots, p),$$

where $g^{(i)}(\theta)$ is the i-th order derivative of $g(\theta)$ with respect to θ. Then

$$\inf_{\hat{g} \in G} V_{\theta_0}(\hat{g}) = \left(g^{(1)}(\theta_0), \ldots, g^{(k)}(\theta_0) \right) \Lambda^{-1} \left(g^{(1)}(\theta_0), \ldots, g^{(k)}(\theta_0) \right)'$$

where Λ is a $k \times k$ matrix whose elements are

$$\lambda_{ij} = \int_{a(\theta_0)}^{b(\theta_0)} \frac{\dfrac{\partial^i}{\partial \theta^i} f(x, \theta_0) \dfrac{\partial^j}{\partial \theta^j} f(x, \theta_0)}{f(x, \theta_0)} d\mu \quad (i, j = 1, \ldots, k).$$

The proof is omitted. In the following theorem we shall get a sharp lower bound.

Theorem 2.3.2. *Let $g(\theta)$ be $(p+1)$-times differentiable over Θ. Let $\hat{g}(X)$ be an unbiased estimator of $g(\theta)$. If for $p \geq 1$ and a fixed θ_0, the conditions (A.2.3.1) to (A.2.3.7) hold, then*

$$\inf_{\hat{g}\,:\,\text{unbiased}} V_{\theta_0}(\hat{g}) = v_k(\theta_0),$$

that is, the bound $v_k(\theta_0)$ is sharp, where

$$v_k(\theta_0) = \left(g^{(1)}(\theta_0), \ldots, g^{(k)}(\theta_0)\right) \Lambda^{-1} \left(g^{(1)}(\theta_0), \ldots, g^{(k)}(\theta_0)\right)'$$

with a $k \times k$ matrix Λ given in Lemma 2.3.1.

Proof. From the unbiasedness condition of $\hat{g}(X)$, (A.2.3.2) and (A.2.3.7) we have

$$(2.3.4) \qquad \int_{a(\theta_0)}^{b(\theta_0)} \hat{g}(x) f(x, \theta_0) d\mu = g(\theta_0),$$

$$(2.3.5) \qquad \int_{a(\theta_0)}^{b(\theta_0)} \hat{g}(x) \left\{ \frac{\partial^i}{\partial \theta^i} f(x, \theta_0) \right\} d\mu = g^{(i)}(\theta_0) \qquad (i = 1, \ldots, p).$$

By Lemma 2.3.1 it follows that the sharp lower bound of

$$V_{\theta_0}(\hat{g}) = \int_{a(\theta_0)}^{b(\theta_0)} \{\hat{g}(x) - g(\theta_0)\}^2 f(x, \theta_0) d\mu(x)$$

under (2.3.4) and (2.3.5) is given by $\left(g^{(1)}(\theta_0), \ldots, g^{(k)}(\theta_0)\right) \Lambda^{-1}\left(g^{(1)}(\theta_0), \ldots, g^{(k)}(\theta_0)\right)'$, i.e.,

$$(2.3.6) \quad \inf_{\hat{g}\,:\,(2.3.4),\,(2.3.5)} V_{\theta_0}(\hat{g}) = \left(g^{(1)}(\theta_0), \ldots, g^{(k)}(\theta_0)\right) \Lambda^{-1} \left(g^{(1)}(\theta_0), \ldots, g^{(k)}(\theta_0)\right)'$$

$$= v_k(\theta_0) \quad \text{(say)}.$$

Note that the right-hand side of (2.3.6) is the Bhattacharyya bound for the variance of unbiased estimators at $\theta = \theta_0$ (see Theorem 2.2.1). From (2.3.6) it follows that for any $\varepsilon > 0$ there exists $\hat{g}_\varepsilon(x)$ in the interval $(a(\theta_0), b(\theta_0))$ satisfying (2.3.4), (2.3.5) and

$$V_{\theta_0}(\hat{g}_\varepsilon) \leq v_k(\theta_0) + \varepsilon.$$

We can extend $\hat{g}_\varepsilon(x)$ for x outside $(a(\theta_0), b(\theta_0))$ from the unbiasedness condition

$$(2.3.7) \qquad \int_{a(\theta)}^{b(\theta)} \hat{g}_\varepsilon(x) f(x, \theta) d\mu(x) = g(\theta) \qquad \text{for all} \quad \theta \in \Theta.$$

For $\theta > \theta_0$, i.e., $b(\theta) \geq b(\theta_0)$ we put

$$h(\theta) = \int_{a(\theta)}^{b(\theta_0)} \hat{g}_\varepsilon(x) f(x, \theta) d\mu(x).$$

By (2.3.7) we obtain

$$(2.3.8) \qquad \int_{b(\theta_0)}^{b(\theta)} \hat{g}_\varepsilon(x) f(x,\theta) d\mu(x) = g(\theta) - h(\theta).$$

Differentiating $(p+1)$-times both sides of (2.3.8) with respect to θ, recursively, we have by (A.2.3.1), (A.2.3.2), (A.2.3.5), (A.2.3.6) and (A.2.3.7)

$$\int_{b(\theta_0)}^{b(\theta)} \hat{g}_\varepsilon(x) \left\{ \frac{\partial}{\partial \theta} f(x,\theta) \right\} d\mu(x) = g^{(1)}(\theta) - h^{(1)}(\theta),$$

$$\int_{b(\theta_0)}^{b(\theta)} \hat{g}_\varepsilon(x) \left\{ \frac{\partial^p}{\partial \theta^p} f(x,\theta) \right\} d\mu(x) = g^{(p)}(\theta) - h^{(p)}(\theta),$$

$$(2.3.9) \quad B_p(\theta) b^{(1)}(\theta) \hat{g}_\varepsilon(b(\theta)) + \int_{b(\theta_0)}^{b(\theta)} \hat{g}_\varepsilon(x) \left\{ \frac{\partial^{p+1}}{\partial \theta^{p+1}} f(x,\theta) \right\} d\mu(x) = g^{(p+1)}(\theta) - h^{(p+1)}(\theta).$$

Differentiation under the integral sign is admitted because of (A.2.3.7). If $\hat{g}_\varepsilon(x)$ satisfies (2.3.9), then it also satisfies (2.3.8) since $g^{(i)}(\theta_0) = h^{(i)}(\theta_0)$ $(i = 1, \ldots, p)$ and $g(\theta_0) = h(\theta_0)$. Note that $h^{(p+1)}(\theta)$ is determined by the values of $\hat{g}_\varepsilon(x)$ for $a(\theta) < x < b(\theta_0)$, where it is already given. Since the integral equation (2.3.9) is of Volterra's second type, it follows that the solution $\hat{g}_\varepsilon(x)$ exists for $b(\theta) > x \geq b(\theta_0)$. Similarly we can construct $\hat{g}_\varepsilon(x)$ for $x < a(\theta_0)$. Repeating the same process we can define $\hat{g}_\varepsilon(x)$ for all x. Hence we have

$$\inf_{\hat{g}\,:\,\text{unbiased}} V_{\theta_0}(\hat{g}) = v_k(\theta_0)$$

i.e., $v_k(\theta_0)$ is a sharp bound. Thus we have completed the proof.

We shall give one example corresponding to each of the situations where the conditions (A.2.3.1), (A.2.3.2)/(A.2.3.2)* to (A.2.3.7) are assumed.

Example 2.3.1. Let X be a real random variable with a density function $f(x,\theta)$ (with respect to the Lebesgue measure μ) satisfying each case.

(i) Location parameter case. The density function is of the form $f_0(x - \theta)$ and satisfies the following :

$$f_0(x) > 0 \qquad \text{for} \quad a < x < b,$$
$$f_0(x) = 0 \qquad \text{for} \quad x \leq a,\ x \geq b,$$

and $\lim_{x \to a+0} f_0(x) > 0$, $\lim_{x \to b-0} f_0(x) > 0$ and $f_0(x)$ is continuously differentiable in the open interval (a, b).

(ii) The case on estimation of $g(\theta) = \theta$ with a density function

$$f_0(x - \theta) = \begin{cases} c(1 - (x - \theta)^2)^{q-1} & \text{for} \quad |x - \theta| < 1, \\ 0 & \text{for} \quad |x - \theta| \geq 1 \end{cases}$$

is discussed in Section 2.2, where $q > 1$ and c is some constant.

(iii) Scale parameter case. The density function of the form $f_1(x/\theta)/\theta$ satisfies the following :

$$f_1(x) > 0 \qquad \text{for} \quad 0 < a < x < b,$$
$$f_1(x) = 0 \qquad \text{otherwise},$$

and satisfies the same condition in (i).

(iv) Selection parameter case (see, e.g., Morimoto and Sibuya (1967), Zacks (1971)). Consider a family of density functions whose supporting intervals depend on a selection parameter θ and are of the form $(\theta, b(\theta))$, where $-\infty < \theta < b(\theta) < \infty$ and $b(\theta)$ is a non-decreasing function of θ and almost everywhere differentiable. Such a family of density functions is specified by

$$f(x, \theta) = \begin{cases} \dfrac{p_0(x)}{F(\theta)} & \text{for} \quad \theta \leq x \leq b(\theta), \\ 0 & \text{otherwise}, \end{cases}$$

where $p_0(x) > 0$ a.e. and $F(\theta) = \int_\theta^{b(\theta)} p_0(x)d\mu(x)$. Note that the cases (i), (iii) and (iv) correspond to the case $p = 0$, and (ii) corresponds to the case $p = q - 1$, where q is an integer.

The lower bound for the variance of unbiased estimators for a sample of size n of real-valued observations

Now suppose that we have a sample of size n, $(X_1, \ldots, X_n)$ of which X_i's are independently and identically distributed according to the distribution characterized in the above. Then we can define statistics

$$Y = \left(\max_{1 \leq i \leq n} X_i + \min_{1 \leq i \leq n} X_i \right) \Big/ 2, \quad Z = \left(\max_{1 \leq i \leq n} X_i - \min_{1 \leq i \leq n} X_i \right) \Big/ 2$$

and we may concentrate our attention on the estimators depending only on Y and Z. Since they are not sufficient statistics, we may lose some information by doing so. More generally, for a sample of size n, $(X_1, \ldots, X_n)$ from a population in a one-directional family of densities with a support $A(\theta)$, we can define two statistics

$$\overline{\theta} = \sup \left\{ \theta \,|\, X_i \in A(\theta) \; (i = 1, \ldots, n) \right\},$$
$$\underline{\theta} = \inf \left\{ \theta \,|\, X_i \in A(\theta) \; (i = 1, \ldots, n) \right\},$$

and also define

$$Y = \frac{1}{2} \left(\overline{\theta} + \underline{\theta} \right), \qquad Z = \frac{1}{2} \left(\overline{\theta} - \underline{\theta} \right).$$

There are various ways of defining the pair of statistics Y and Z, but disregarding their construction we assume that there exists a pair (Y, Z) which satisfies the following.

Let Y and Z be real-valued statistics based on a sample $(X_1, \ldots, X_n)$ of size n for $n \geq 2$. We assume that (Y, Z) has a joint probability density function $f_\theta(y, z)$ (with respect to the Lebesgue measure $\mu_{y,z}$) satisfying

$$f_\theta(y, z) = f_\theta(y|z)h_\theta(z), \qquad \text{a.e.},$$

where $f_\theta(y|z)$ is a conditional density function of y given z with respect to the Lebesgue measure μ_y and $h_\theta(z)$ is a density function of z with respect to the Lebesgue measure μ_z. Note that if Z is ancillary, $h_\theta(z)$ is independent of θ. We assume the following condition :

(A.2.3.1)' For almost all z $[\mu_z]$

$$f_\theta(y|z) > 0 \quad \text{for} \quad a_z(\theta) < y < b_z(\theta),$$
$$f_\theta(y|z) = 0 \quad \text{for} \quad y \le a_z(\theta),\ y \ge b_z(\theta),$$

where $a_z(\theta)$ and $b_z(\theta)$ are strictly monotone increasing functions of θ for almost all z $[\mu_z]$ which depend on z, and

$$h_\theta(z) > 0 \quad \text{for} \quad c < z < d,$$
$$h_\theta(z) = 0 \quad \text{for} \quad z \le c,\ z \ge d,$$

where c and d are constants independent of θ. We also assume that, for almost all z $[\mu_z]$, $f_\theta(y|z)$ instead of $f(x,\theta)$ satisfies the conditions (A.2.3.2) to (A.2.3.7), and we call the corresponding conditions (A.2.3.2)' to (A.2.3.7)'.

Let $\hat{g}(Y,Z)$ be any unbiased estimator of $g(\theta)$. We define

$$(2.3.10) \qquad \phi_z(\theta) = \int_{a_z(\theta)}^{b_z(\theta)} \hat{g}(y,z) f_\theta(y|z) d\mu_y \qquad \text{for} \quad a.a.\ z\,[\mu_z].$$

Further we assume the following condition :

$$(A.2.3.8) \qquad \phi_z(\theta_0) = g(\theta_0) \qquad \text{for} \quad a.a.\ z\,[\mu_z],$$
$$\phi_z^{(i)}(\theta_0) = 0 \qquad \text{for} \quad a.a.\ z\,[\mu_z] \quad (i = 1,\ldots,k),$$

where $\phi_z^{(i)}(\theta)$ is the i-th order derivative of $\phi_z(\theta)$ with respect to θ.

In the following theorem we shall show that the sharp bound is equal to zero.

Theorem 2.3.3. Let $g(\theta)$ be $(p+1)$-times differentiable over Θ. Let $\hat{g}(X_1,\ldots,X_n)$ be an unbiased estimator of $g(\theta)$. If $n \ge 2$ and, for a fixed θ_0, the conditions (A.2.3.1)' to (A.2.3.7)' and (A.2.3.8) hold, then,

$$\inf_{\hat{g}\,:\,\text{unbiased}} V_{\theta_0}(\hat{g}) = 0.$$

Proof. From the unbiasedness condition we obtain

$$(2.3.11) \qquad \int_c^d \phi_z(\theta) h_\theta(z) d\mu_z = g(\theta) \qquad \text{for all} \quad \theta \in \Theta.$$

First we assume that $\phi_z(\theta)$ is given, then under (A.2.3.1)' to (A.2.3.7)' we have by Lemma 2.3.1

$$(2.3.12) \quad \inf_{\hat{g}\,:\,(2.3.10)} \int_{a_z(\theta_0)}^{b_z(\theta_0)} \{\hat{g}(y,z) - \phi_z(\theta_0)\}^2 f_{\theta_0}(y|z) d\mu_y$$
$$= \left(\phi_z^{(1)}(\theta_0),\ldots,\phi_z^{(k)}(\theta_0) \right) \Lambda^{-1} \left(\phi_z^{(1)}(\theta_0),\ldots,\phi_z^{(k)}(\theta_0) \right)'$$

$$= v_k(\theta_0|z) \quad \text{(say)},$$

where $\phi_z^{(i)}(\theta)$ is the i-th order derivative of $\phi_z(\theta)$ with respect to θ and Λ is a $k \times k$ matrix whose elements are

$$\lambda_{ij} = \int_{a_z(\theta_0)}^{b_z(\theta_0)} \frac{1}{f_{\theta_0}(y|z)} \left\{ \frac{\partial^i}{\partial\theta^i} f_{\theta_0}(y|z) \right\} \left\{ \frac{\partial^j}{\partial\theta^j} f_{\theta_0}(y|z) \right\} d\mu_y.$$

From (2.3.12) it follows that for any $\varepsilon > 0$ there exists $\hat{g}_\varepsilon(y,z)$ such that

(2.3.13)
$$\int_{a_z(\theta)}^{b_z(\theta)} \hat{g}_\varepsilon(y,z) f_\theta(y|z) d\mu_y = \phi_z(\theta) \qquad \text{for all} \quad \theta \in \Theta,$$

(2.3.14)
$$\int_{a_z(\theta_0)}^{b_z(\theta_0)} \{\hat{g}_\varepsilon(y,z) - \phi_z(\theta_0)\}^2 f_{\theta_0}(y|z) d\mu_y < v_k(\theta_0|z) + \varepsilon.$$

Since by (2.3.11)

$$\int_c^d \int_{a_z(\theta)}^{b_z(\theta)} \hat{g}_\varepsilon(y,z) f_\theta(y|z) h_\theta(z) d\mu_y d\mu_z = g(\theta)$$

for all $\theta \in \Theta$, it follows from (2.3.14) that

(2.3.15)
$$\int_c^d \int_{a_z(\theta_0)}^{b_z(\theta_0)} \{\hat{g}_\varepsilon(y,z) - \phi_z(\theta_0)\}^2 f_{\theta_0}(y|z) h_{\theta_0}(z) d\mu_y d\mu_z$$
$$< \int_c^d v_k(\theta_0|z) h_{\theta_0}(z) d\mu_z + \varepsilon.$$

By the condition (A.2.3.8)

$$v_k(\theta_0|z) = 0 \qquad \text{for} \quad a.a. \ z \ [\mu_z].$$

From (2.3.15) we obtain

(2.3.16)
$$\int_c^d \int_{a_z(\theta_0)}^{b_z(\theta_0)} \{\hat{g}_\varepsilon(y,z) - \phi_z(\theta_0)\}^2 f_{\theta_0}(y|z) h_{\theta_0}(z) d\mu_y d\mu_z < \varepsilon.$$

Putting $\hat{g}_\varepsilon(x_1,\ldots,x_n) = \hat{g}_\varepsilon(y,z)$, we have by (2.3.13) and (2.3.16)

$$E_\theta(\hat{g}_\varepsilon) = g(\theta) \qquad \text{for all} \quad \theta \in \Theta,$$
$$V_{\theta_0}(\hat{g}_\varepsilon) < \varepsilon.$$

Letting $\varepsilon \to 0$ we obtain

$$\inf_{\hat{g}\,:\,\text{unbiased}} V_{\theta_0}(\hat{g}) = 0.$$

Thus we complete the proof.

Now we shall find a function $\phi_z(\theta)$ satisfying the condition (A.2.3.8) when $\Theta = \mathbf{R}^1$, $g(\theta) = \theta$ and μ_z is a Lebesgue measure. Without loss of generality, we put $\theta_0 = 0$. We define

$$(2.3.17) \qquad \phi_z(\theta) = \frac{M\,\mathrm{sgn}\theta|\theta|^{k+2}(z-c)^k}{h_\theta(z)\{(d-z)^{k+2} + |\theta|^{k+2}(z-c)^{k+2}\}},$$

where M is a constant and k is a positive integer. Then we have

$$
\begin{aligned}
\int_c^d \phi_z(\theta)h_\theta(z)dz &= M \int_c^d \frac{\mathrm{sgn}\theta|\theta|^{k+2}(z-c)^k}{(d-z)^{k+2} + |\theta|^{k+2}(z-c)^{k+2}}dz \\
&= \frac{M}{d-c}\mathrm{sgn}\theta|\theta|^{k+2}\int_0^\infty \frac{u^k}{1+|\theta|^{k+2}u^{k+2}}du \\
&\quad \left(\text{after transformation } u = \frac{z-c}{d-z}\right) \\
&= \frac{M}{d-c}\mathrm{sgn}\theta|\theta|\int_0^\infty \frac{v^k}{1+v^{k+2}}dv = \frac{MK}{d-c}\theta,
\end{aligned}
$$

where $K = \displaystyle\int_0^\infty \frac{v^k}{1+v^{k+2}}dv$ is a constant. If we put $M = (d-c)/K$, then

$$\int_c^d \phi_z(\theta)h_\theta(z)dz = \theta.$$

And it is easily seen that

$$
\begin{aligned}
\phi_z(0) &= 0 &&\text{for } a.a.\ z, \\
\phi_z^{(i)}(0) &= 0 &&\text{for } a.a.\ z \quad (i = 1, \ldots, k).
\end{aligned}
$$

Thus it is shown that $\phi_z(\theta)$ given by (2.3.17) satisfies the condition (A.2.3.8).

We consider the estimation on the location parameter θ. Let X_1 and X_2 be independent and identically distributed random variables with a density function $f(x, \theta)$ of the form $f_0(x - \theta)$ which satisfies the following :

$$
\begin{aligned}
f_0(x) &> 0 &&\text{for } a < x < b, \\
f_0(x) &= 0 &&\text{for } x \le a,\ x \ge b,
\end{aligned}
$$

and $\lim_{x\to a+0} f_0(x) > 0$, $\lim_{x\to b-0} f_0(x) > 0$ and $f_0(x)$ is continuously differentiable in the open interval (a, b). We define

$$Y = \frac{1}{2}(X_1 + X_2), \quad Z = \frac{1}{2}(X_1 - X_2).$$

Then if the conditions (A.2.3.1)$'$ to (A.2.3.7)$'$ and (A.2.2.8) are assumed, we have

$$\inf_{\hat\theta:\,\text{unbiased}} V_{\theta_0}\left(\hat\theta\right) = 0.$$

Note that Z is an ancillary statistic, but that (Y, Z) is not sufficient unless $f_0(x)$ is constant for $a < x < b$.

2.4. A SECOND TYPE OF ONE-DIRECTIONAL DISTRIBUTION AND THE LOWER BOUND FOR THE VARIANCE OF UNBIASED ESTIMATORS

Suppose that (X, Y) is a pair of real random variables according to a joint density function $f(x, y, \theta)$ (with respect to the Lebesgue measure μ) which has the product set $(0, a(\theta)) \times (0, b(\theta))$ of two open intervals as its support $A(\theta)$, where $a(\theta)$ is a monotone increasing function and $b(\theta)$ is a monotone decreasing function. We assume the condition

$$\text{(A.2.4.1)} \qquad \inf_{(x,y) \in A(\theta)} f(x, y, \theta) > 0.$$

Let the marginal density functions of X and Y be $f_1(x, \theta)$ and $f_2(y, \theta)$ with respect to the Lebesgue measures μ_x and μ_y, respectively. We further make the following assumption:
(A.2.4.2) The density functions $f_1(x, \theta)$ and $f_2(y, \theta)$ are continuously differentiable in θ and satisfy the conditions (A.2.3.5), (A.2.3.6) and (A.2.3.7) for $k = 1$ when $f_1(x, \theta)$ and $f_2(y, \theta)$ are substituted instead of $f(x, \theta)$ in them.
In the following theorem we shall show that the sharp bound is equal to zero.

Theorem 2.4.1. Let $\Theta = \mathbf{R}^1$. Suppose that X and Y are random variables with a joint density function $f(x, y, \theta)$ (with respect to the Lebesgue measure μ) satisfying (A.2.4.1) and (A.2.4.2) for a fixed θ_0. Let $g(\theta)$ be continuously differentiable over Θ. Let $\hat{g}(X, Y)$ be an unbiased estimator of $g(\theta)$. Then

$$\inf_{\hat{g}:\text{unbiased}} V_{\theta_0}(\hat{g}) = 0.$$

Proof. We first define

$$\hat{g}(x, y) = g(\theta_0) \qquad \text{for} \quad (x, y) \in A(\theta_0).$$

In order to extend $\hat{g}(x, y)$ for (x, y) outside $A(\theta_0)$ using the unbiasedness condition, we consider unbiased estimators $\hat{g}_1(X)$ and $\hat{g}_2(Y)$ of $g(\theta)$ with respect to $f_1(x, \theta)$ and $f_2(y, \theta)$, respectively, such that $\hat{g}_1(x) = g(\theta_0)$ for $0 < x < a(\theta_0)$ and $\hat{g}_2(y) = g(\theta_0)$ for $0 < y < b(\theta_0)$. For $\theta > \theta_0$, i.e., $a(\theta) \geq a(\theta_0)$ we put

$$h_1(\theta) = g(\theta_0) \int_0^{a(\theta_0)} f_1(x, \theta) d\mu_x.$$

and also for $\theta \leq \theta_0$, i.e., $b(\theta) \geq b(\theta_0)$

$$h_2(\theta) = g(\theta_0) \int_0^{b(\theta_0)} f_2(y, \theta) d\mu_y.$$

Since $\hat{g}_1(X)$ and $\hat{g}_2(Y)$ are unbiased estimator of $g(\theta)$, it follows that

(2.4.1)
$$\int_{a(\theta_0)}^{a(\theta)} \hat{g}_1(x)f_1(x,\theta)d\mu_x = g(\theta) - h_1(\theta) \qquad \text{for all} \quad \theta \geq \theta_0,$$

$$\int_{b(\theta_0)}^{b(\theta)} \hat{g}_2(y)f_2(y,\theta)d\mu_y = g(\theta) - h_2(\theta) \qquad \text{for all} \quad \theta \leq \theta_0.$$

Since the supports of the density functions $f_1(x,\theta)$ and $f_2(y,\theta)$ are open intervals $(0, a(\theta))$ and $(0, b(\theta))$, respectively, it follows from (A.2.4.1) that

(2.4.2)
$$0 < \lim_{x \to a(\theta)-0} f_1(x,\theta) = \alpha_1(\theta) \qquad \text{(say)},$$

$$0 < \lim_{y \to b(\theta)-0} f_2(y,\theta).$$

Differentiating both sides of (2.4.1) we have by (A.2.4.2) and (2.4.2)

(2.4.3)
$$\hat{g}_1(a(\theta))\alpha_1(\theta) + \int_{a(\theta_0)}^{a(\theta)} \hat{g}_1(x)\left\{\frac{\partial}{\partial\theta}f_1(x,\theta)\right\}d\mu_x = g'(\theta) - h_1'(\theta)$$

for all $\theta \geq \theta_0$. Since the equation (2.4.3) is of Volterra's second type, it follows that the solution $\hat{g}_1(x)$ exists for all $x \geq a(\theta_0)$. If $\hat{g}_1(x)$ satisfies (2.4.3), then it also satisfies (2.4.1) since $g(\theta_0) = h_1(\theta_0)$. Similarly we can construct the unbiased estimator $\hat{g}_2(y)$ for all $y \geq b(\theta_0)$. We define an estimator

(2.4.4)
$$\hat{g}(x,y) = \begin{cases} g(\theta_0) & \text{for} \quad 0 < x < a(\theta_0),\ 0 < y < b(\theta_0), \\ \hat{g}_1(x) & \text{for} \quad a(\theta_0) \leq x,\ 0 < y < b(\theta_0), \\ \hat{g}_2(y) & \text{for} \quad 0 < x < a(\theta_0),\ b(\theta_0) \leq y. \end{cases}$$

Then $\hat{g}(X,Y)$ is an unbiased estimator of $g(\theta)$ with variance 0 at $\theta = \theta_0$.

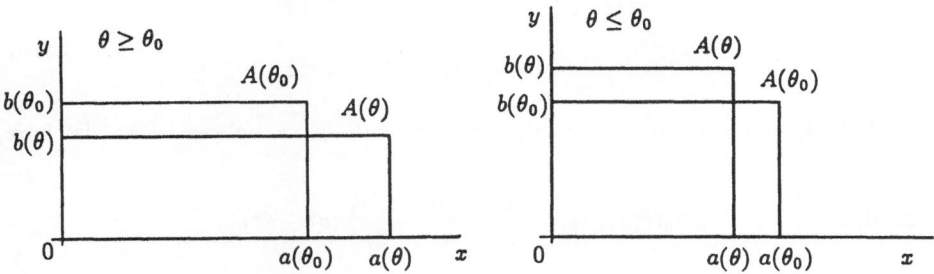

Figure 2.4.1. The supports $A(\theta_0)$ and $A(\theta)$ of the density functions $f(x,y,\theta_0)$ and $f(x,y,\theta)$ for $\theta \geq \theta_0$ and $\theta \leq \theta_0$.

Indeed, we have from (2.4.1) and (2.4.4)

$$E_\theta[\hat{g}(X,Y)] = \int_0^{b(\theta)} \int_0^{a(\theta_0)} g(\theta_0) f(x,y,\theta) d\mu + \int_0^{b(\theta)} \int_{a(\theta_0)}^{a(\theta)} \hat{g}_1(x) f(x,y,\theta) d\mu$$

$$= g(\theta_0) \int_0^{a(\theta_0)} f_1(x,\theta) d\mu_x + \int_{a(\theta_0)}^{a(\theta)} \hat{g}_1(x) f_1(x,\theta) d\mu_x$$

for all $\theta \geq \theta_0$. Similarly we have that $E_\theta[\hat{g}(X,Y)] = g(\theta)$ for all $\theta \leq \theta_0$. Hence we see that $\hat{g}(X,Y)$ is an unbiased estimator of $g(\theta)$. We also have, for all $\theta \geq \theta_0$

$$V_{\theta_0}(\hat{g}(X,Y)) = g^2(\theta_0) \int_0^{a(\theta_0)} f_1(x,\theta) d\mu_x - \int_{a(\theta_0)}^{a(\theta)} \hat{g}_1^2(x) f_1(x,\theta) d\mu_x - g^2(\theta).$$

When $\theta = \theta_0$, we obtain

$$V_{\theta_0}(\hat{g}(X,Y)) = 0.$$

Thus we complete the proof.

We can give the following example.
Example 2.4.1. Let $X_1, \ldots, X_n$ and $Y_1, \ldots, Y_n$ be independent and identically distributed random variables with the uniform distributions $U(0,\theta)$ and $U(0,1/\theta)$, respectively. Put

$$T_1 = \max_{1 \leq i \leq n} X_i, \quad T_2 = \max_{1 \leq i \leq n} X_i.$$

Then the unbiased estimator $\hat{\theta}(T_1, T_2)$ of θ with variance 0 at $\theta = 1$ is given by

(2.4.5) $\qquad \hat{\theta}(t_1, t_2) = \begin{cases} \hat{\theta}_1(t_1) & \text{for } 1 \leq t_1, \ 0 < t_2 < 1, \\ \hat{\theta}_2(t_2) & \text{for } 1 \leq t_2, \ 0 < t_1 < 1, \\ 1 & \text{for } t_1 < 1, \ t_2 < 1, \end{cases}$

where

$$\hat{\theta}_1(t_1) = \left(1 + \frac{1}{n}\right) t_1 \qquad \text{for } 1 \leq t_1,$$

$$\hat{\theta}_2(t_2) = \left(1 - \frac{1}{n}\right) t_2 \qquad \text{for } 1 \leq t_2.$$

Indeed, we can easily see that the estimator $\hat{\theta}(T_1, T_2)$ is unbiased. We also have, for $0 < \theta \leq 1$

(2.4.6) $\qquad V_\theta(\hat{\theta}(T_1, T_2)) = \left\{1 - \frac{(n-1)^2}{n(n-2)}\right\} (\theta^n - \theta^2).$

We also have for $\theta \geq 1$

(2.4.7) $\qquad V_\theta(\hat{\theta}(T_1, T_2)) = \left\{1 - \frac{(n+1)^2}{n(n-2)}\right\} (\theta^{-n} - \theta^2).$

From (2.4.6) and (2.4.7), we obtain

$$V_1(\hat{\theta}(T_1, T_2)) = 0.$$

Thus we see that the unbiased estimator $\hat{\theta}(T_1, T_2)$ given by (2.4.5) has variance 0 at $\theta = 1$.

As a further case of this situation, let X and Y be independent real random variables according to density functions $(1/\theta)f_1(x/\theta)$ and $\theta f_1(\theta y)$ (with respect to the Lebesgue measure μ) with a positive valued parameter θ, respectively, which satisfy the following:

(A.2.4.3)
$$f_1(x) > 0 \quad \text{for} \quad 0 < x < 1,$$
$$f_1(x) = 0 \quad \text{otherwise,}$$

and $f_1(x)$ is $(p+1)$-times continuously differentiable in the open interval $(0, 1)$ and for each $i = 0, 1, \ldots, p$

$$0 < \lim_{x \to 0+0} \left| \frac{f_1^{(i)}(x)}{x^{p-i}} \right| < \infty, \quad 0 < \lim_{x \to 1-0} \left| \frac{f_1^{(i)}(x)}{(1-x)^{p-i}} \right| < \infty.$$

By Theorem 2.3.3 we have the following:

Theorem 2.4.2. *Let $g(\theta)$ be an estimable function which is $(p+1)$-times differentiable over $\mathbf{R}^1$. Let $\hat{g}(X, Y)$ be an unbiased estimator of $g(\theta)$. If the conditions (A.2.4.3) and (A.2.3.8) on $p_\theta(x|t)$ hold, then*

$$\inf_{\hat{g}\,:\,\text{unbiased}} V_{\theta_0}(\hat{g}) = 0,$$

where $p_\theta(x|t)$ denotes the conditional density function of X given $XY = t$.

Proof. Letting $T = XY$, we have the conditional density function $p_\theta(x|t)$ of X given $T = t$:

(2.4.8)
$$p_\theta(x|t) = \begin{cases} \dfrac{\dfrac{1}{x}f_1\left(\dfrac{x}{\theta}\right)f_1\left(\dfrac{\theta t}{x}\right)}{\displaystyle\int_{\theta t}^{\theta} \dfrac{1}{x}f_1\left(\dfrac{x}{\theta}\right)f_1\left(\dfrac{\theta t}{x}\right)dx} & \text{for} \quad \theta t < x < \theta, \\[6pt] 0 & \text{otherwise,} \end{cases}$$

for almost all $t[\mu]$. Since for $\theta = 1$ and almost all $t[\mu]$

$$p_1(x|t) = \begin{cases} \dfrac{c_t}{x}f_1(x)f_1\left(\dfrac{t}{x}\right) & \text{for} \quad t < x < 1, \\[6pt] 0 & \text{otherwise,} \end{cases}$$

we obtain from (2.4.8)

$$p_\theta(x|t) = \frac{1}{\theta}p_1\left(\frac{x}{\theta}\Big|t\right) \quad \text{for all} \quad \theta, \quad a.a. \ t\,[\mu]$$

where

$$c_t = \left(\int_t^1 \frac{1}{x} f_1(x) f_1\left(\frac{t}{x}\right) dx \right)^{-1}.$$

Putting

$$g_t(x) = \begin{cases} \dfrac{c_t}{x} f_1\left(\dfrac{t}{x}\right) & \text{for } t < x < 1, \\ 0 & \text{otherwise,} \end{cases}$$

for almost all t $[\mu]$, we have

(2.4.9) $p_1(x|t) = f_1(x) g_t(x), \qquad a.a. \ t \ [\mu],$

hence the same conditions on $p_1(x|t)$ as (A.2.3.1)' and (A.2.3.2)' hold. Indeed, we have by (A.2.4.3) and (2.4.9)

(2.4.10)
$$\lim_{x \to t+0} p_1(x|t) = f_1(t+0) g_t(t+0) = \frac{c_t}{t} f_1(t+0) f_1(1+0) = 0,$$
$$\lim_{x \to 1-0} p_1(x|t) = f_1(1-0) g_t(1-0) = 0$$

for almost all t $[\mu]$. By the Leibniz formula, we obtain

(2.4.11)
$$\frac{\partial^i p_1(x|t)}{\partial x^i} = \sum_{j=0}^i \binom{i}{j} f_1^{(j)}(x) g_t^{(i-j)}(x)$$

$$= \sum_{j=0}^i \binom{i}{j} f_1^{(i-j)}(x) g_t^{(j)}(x), \qquad a.a. \ t \ [\mu].$$

By (A.2.4.3) we have for $i = 1, \dots, p-1$ and almost all t $[\mu]$

(2.4.12) $\displaystyle \lim_{x \to 1-0} \frac{\partial^i p_1(x|t)}{\partial x^i} = 0, \quad \lim_{x \to 1-0} \frac{\partial^p p_1(x|t)}{\partial x^p} = C_t \neq 0$

where C_t is finite. Since

$$g_t^{(i)}(x) = c_t \sum_{j=0}^i (-1)^j j! \frac{1}{x^{j+1}} \frac{\partial^{j-i}}{\partial x^{j-i}} f_1\left(\frac{t}{x}\right),$$

it follows by (A.2.4.3) and (2.4.11) that for almost all t $[\mu]$,

(2.4.13)
$$\lim_{x \to t+0} \frac{\partial^i p_1(x|t)}{\partial x^i} = 0 \qquad (i = 1, \dots, p-1),$$
$$\lim_{x \to t+0} \frac{\partial^p p_1(x|t)}{\partial x^p} = D_t \neq 0,$$

where D_t is finite. It is seen by (2.4.9), (2.4.10), (2.4.12) and (2.4.13) that the same condition on $p_1(x|t)$ as (A.2.3.2)' holds.

For $i = 1, \ldots, [p/2]$

$$0 < \int_t^1 \frac{\left\{ \dfrac{\partial^i}{\partial x^i} p_1(x|t) \right\}^2}{p_1(x|t)} dx < \infty, \qquad a.a.\ t\ [\mu]$$

and

$$\int_t^1 \frac{\left\{ \sum_{i=[p/2]+1}^p c_i \dfrac{\partial^i}{\partial x^i} p_1(x|t) \right\}^2}{p_1(x|t)} dx$$

is finite for *a.a.* t $[\mu]$ unless $c_{[p/2]+1} = \cdots = c_p$, where $[s]$ denotes the largest integer less than or equal to s, since when $x \to 0 + 0$ or $x \to t - 0$ the numerator of the integrand approaches to a polynomial in x or $t - x$ of the degree $p - i^*$ if $c_{i^*} \neq 0$ and $c_{i^*+1} = \cdots = c_p = 0$ and the denominator tends to that of the degree p. Hence the same condition on $p_1(x|t)$ as (A.2.3.4)' holds for $k = [p/2]$. And also from (A.2.4.1) it is seen that when $x \to 0 + 0$ or $x \to t - 0$, $\{(\partial^i/\partial x^i)p_1(x|t)\}/p_1(x|t)$ $(i = 0, 1, \ldots, k)$ approach to polynomials of different degrees, hence they are linearly independent. Putting

$$\rho_t(x) = \sup_{x' : |x' - x| < c} p_1(x'|t)$$

for appropriate $c > 0$, the same conditions on $p_1(x|t)$ as (A.2.3.5)' to (A.2.3.7)' hold. By Theorem 2.3.3 we obtain the conclusion of Theorem 2.4.2.

When $g(\theta) = \theta$, Example 2.4.1 can be also an example of Theorem 2.4.2.

2.5. LOCALLY MINIMUM VARIANCE UNBIASED ESTIMATION

In the theory of minimum variance unbiased estimation for regular cases, it is usually assumed that the density function is continuous with respect to the unknown parameter θ to be estimated. Though the Chapman-Robbins(1951)-Kiefer(1952) type discussion does not require the continuity of the density function in θ, their bounds are not usually sharp and the precise form of the locally minimum variance unbiased (LMVU) estimator is not given.

In the section we shall give an example where the density function is discontinuous in θ, and obtain the exact forms of the LMVU estimators of the parameter and their variances, and illustrate some aspects of the LMVU estimators in such a situation.

Suppose that the density function $f(x, \theta)$ (with respect to a σ-finite measure μ) is not continuous in θ, while the support $A = \{x | f(x, \theta) > 0\}$ can be defined to be independent of θ. Note that this condition does not affect the existence of a LMVU estimator, because it can be shown that a LMVU estimator at $\theta = \theta_0$ exists if

$$\int_A \frac{\{f(x, \theta)\}^2}{f(x, \theta_0)} d\mu(x) < \infty$$

(see, e.g., Barankin, 1949 and Stein, 1950). But the LMVU estimator sometimes behaves very strangely.

Let $X_1, \ldots, X_n$ be independent and identically distributed random variables according to the following density function with respect to the Lebesgue measure :

$$(2.5.1) \qquad f(x, \theta) = \begin{cases} p & \text{for} \quad 0 \le x \le \theta \quad \text{and} \quad \theta + 1 \le x \le 2 \, ; \\ q & \text{for} \quad \theta < x < \theta + 1 \, ; \\ 0 & \text{otherwise}, \end{cases}$$

where the parameter has the range $0 \le \theta \le 1$ and p and q with $0 < p < q$ and $p + q = 1$ are fixed constants. Since the support of $f(x, \theta)$ is the interval $[0, 2]$, in the subsequent discussion it is enough to consider it as the domain of x.

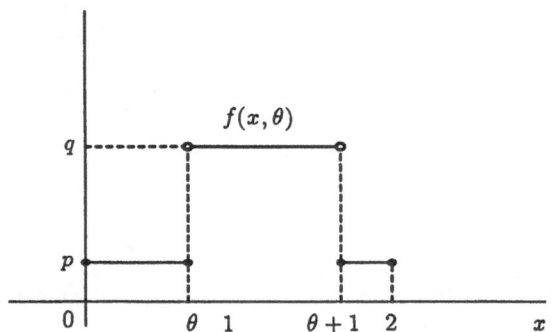

Figure 2.5.1. The density function $f(x, \theta)$ given by (2.5.1).

Then the LMVU estimator $\hat{\theta}_n^* := \hat{\theta}_n^*(x_1, \ldots, x_n)$ at $\theta = \theta_0$ has the form

$$(2.5.2) \qquad \hat{\theta}_n^*(x_1, \ldots, x_n) = \int_0^1 \prod_{i=1}^n \frac{f(x_i, \eta)}{f(x_i, \theta_0)} dG_n(\eta),$$

where G_n is some signed measure over the closed interval $[0, 1]$ (see Stein, 1950 and Section 1.5).

Case I : $n = 1$. We shall construct the LMVU estimator at $\theta = \theta_0$ given by the form (2.5.2) in the case when $n = 1$, i.e., $x_1 = x$. We put

$$(2.5.3) \qquad K_{\theta_0}(\theta, \eta) = \int_0^2 \pi_{\theta_0, \eta}(x) f(x, \theta) dx$$

with

$$\pi_{\theta_0, \eta}(x) = \frac{f(x, \eta)}{f(x, \theta_0)}.$$

In order to obtain the LMVU estimator $\hat{\theta}_1^*(X)$ at $\theta = \theta_0$ it is enough to take some signed measure G over $[0, 1]$ such that

$$E_\theta\left[\hat{\theta}_1^*(X)\right] = \int_0^2 \left\{\int_0^1 \frac{f(x,\eta)}{f(x,\theta_0)} dG(\eta)\right\} f(x,\theta) dx$$

$$= \int_0^1 K_{\theta_0}(\theta,\eta) dG_{(\eta)}$$

$$= \theta$$

for all $\theta \in [0,1]$. For $\eta > \theta_0$ we have

(2.5.4)
$$\pi_{\theta_0,\eta}(x) = \begin{cases} p/q & \text{for} \quad \theta_0 < x \le \eta ; \\ q/p & \text{for} \quad \theta_0 + 1 \le x < \eta + 1 ; \\ 1 & \text{otherwise.} \end{cases}$$

For $\eta < \theta_0$ we also obtain

(2.5.5)
$$\pi_{\theta_0,\eta}(x) = \begin{cases} q/p & \text{for} \quad \eta < x \le \theta_0 ; \\ p/q & \text{for} \quad \eta + 1 \le x < \theta_0 + 1 ; \\ 1 & \text{otherwise.} \end{cases}$$

Note that for $\eta = \theta_0$, $\pi_{\theta_0,\eta}(x) = 1$. In order to calculate $K_{\theta_0}(\theta,\eta)$ we consider two cases $\theta > \theta_0$ and $\theta < \theta_0$. Note that for $\theta = \theta_0$, $K_{\theta_0}(\theta_0,\eta) = 0$.

(i) $\theta > \theta_0$. If $\eta \le \theta_0$, then we have by (2.5.5)

$$K_{\theta_0}(\theta,\eta) = \int_0^2 \pi_{\theta_0,\eta}(x) f(x,\theta) dx$$

$$= \int_0^\eta p\,dx + \int_\eta^{\theta_0} q\,dx + \int_{\theta_0}^\theta p\,dx + \int_\theta^{\eta+1} q\,dx$$

$$+ \int_{\eta+1}^{\theta_0+1} p\,dx + \int_{\theta_0+1}^{\theta+1} q\,dx + \int_{\theta+1}^2 p\,dx$$

$$= 1.$$

If $\theta_0 < \eta < \theta$, then we obtain by (2.5.4)

$$K_{\theta_0}(\theta,\eta) = \int_0^2 \pi_{\theta_0,\eta}(x) f(x,\theta) dx$$

$$= \int_0^{\theta_0} p\,dx + \int_{\theta_0}^\eta \frac{p^2}{q} dx + \int_\eta^\theta p\,dx + \int_\theta^{\theta_0+1} q\,dx$$

$$+ \int_{\theta_0+1}^{\eta+1} \frac{q^2}{p} dx + \int_{\eta+1}^{\theta+1} q\,dx + \int_{\theta+1}^2 p\,dx$$

$$= 1 + c(\eta - \theta_0),$$

where $c = \dfrac{1}{pq} - 4 \ (> 0)$. If $\theta \le \eta$, then we have by (2.5.4)

$$
\begin{aligned}
K_{\theta_0}(\theta, \eta) &= \int_0^2 \pi_{\theta_0, \eta}(x) f(x, \theta) dx \\
&= \int_0^{\theta_0} p\, dx + \int_{\theta_0}^{\theta} \frac{p^2}{q} dx + \int_{\theta}^{\eta} p\, dx + \int_{\eta}^{\theta_0 + 1} q\, dx \\
&\quad + \int_{\theta_0 + 1}^{\theta + 1} \frac{q^2}{p} dx + \int_{\theta + 1}^{\eta + 1} q\, dx + \int_{\eta + 1}^{2} p\, dx \\
&= 1 + c(\theta - \theta_0).
\end{aligned}
$$

Hence we obtain for $\theta > \theta_0$

$$
(2.5.6) \qquad K_{\theta_0}(\theta, \eta) = \begin{cases} 1 & \text{for} \quad \eta \le \theta_0 \ ; \\ 1 + c(\eta - \theta_0) & \text{for} \quad \theta_0 < \eta < \theta \ ; \\ 1 + c(\theta - \theta_0) & \text{for} \quad \theta \le \eta, \end{cases}
$$

where $c = \dfrac{1}{pq} - 4$.

(ii) $\theta < \theta_0$. In a similar way to (i) we have by (2.5.4) and (2.5.5)

$$
(2.5.7) \qquad K_{\theta_0}(\theta, \eta) = \begin{cases} 1 & \text{for} \quad \theta_0 \le \eta \ ; \\ 1 + c(\theta_0 - \eta) & \text{for} \quad \theta < \eta < \theta_0 \ ; \\ 1 + c(\theta_0 - \theta) & \text{for} \quad \eta \le \theta. \end{cases}
$$

We take a signed measure G^* over $[0, 1]$ satisfying $G^*(\{0\}) = -1/c$, $G^*(\{1\}) = 1/c$, $G^*(\{\theta_0\}) = \theta_0$ and $G^*((0, \theta_0) \cup (\theta_0, 1)) = 0$. Since by (2.5.6) and (2.5.7)

$$
K_{\theta_0}(\theta, 0) = \begin{cases} 1 & \text{for} \quad \theta > \theta_0 \ ; \\ 1 + c(\theta_0 - \theta) & \text{for} \quad \theta < \theta_0, \end{cases}
$$

and

$$
K_{\theta_0}(\theta, 1) = \begin{cases} 1 + c(\theta - \theta_0) & \text{for} \quad \theta > \theta_0 \ ; \\ 1 & \text{for} \quad \theta < \theta_0, \end{cases}
$$

we obtain

$$
\begin{aligned}
E_\theta\left[\hat{\theta}_1^*(X)\right] &= \int_0^2 \left\{ \int_0^1 \pi_{\theta_0, \eta}(x) dG^*(\eta) \right\} f(x, \theta) dx \\
&= \int_0^1 K_{\theta_0}(\theta, \eta) dG^*(\eta) \\
&= \frac{1}{c} \left\{ K_{\theta_0}(\theta, 1) - K_{\theta_0}(\theta, 0) \right\} + K_{\theta_0}(\theta, \theta_0) \theta_0 \\
&= \theta
\end{aligned}
$$

for all $\theta \in [0,1]$, which shows the unbiasedness. Hence the LMVU estimator $\hat{\theta}_1^*(x)$ at $\theta = \theta_0$ is given by

$$\hat{\theta}_1^*(x) = \int_0^1 \pi_{\theta_0,\eta}(x) dG^*(\eta)$$

$$= \frac{1}{c} \{\pi_{\theta_0,1}(x) - \pi_{\theta_0,0}(x)\} + \theta_0,$$

where

$$\pi_{\theta_0,1}(x) - \pi_{\theta_0,0}(x) = \begin{cases} 1 - \dfrac{q}{p} & \text{for} \quad 0 \le x \le \theta_0 \ ; \\[2mm] \dfrac{p}{q} - 1 & \text{for} \quad \theta_0 < x < 1 \ ; \\[2mm] 1 - \dfrac{p}{q} & \text{for} \quad 1 < x < \theta_0 + 1 \ ; \\[2mm] \dfrac{q}{p} - 1 & \text{for} \quad \theta_0 + 1 \le x \le 2. \end{cases}$$

Note that $\hat{\theta}_1^*(x)$ can take values outside of the interval $[0,1]$ which is the range of the parameter θ.

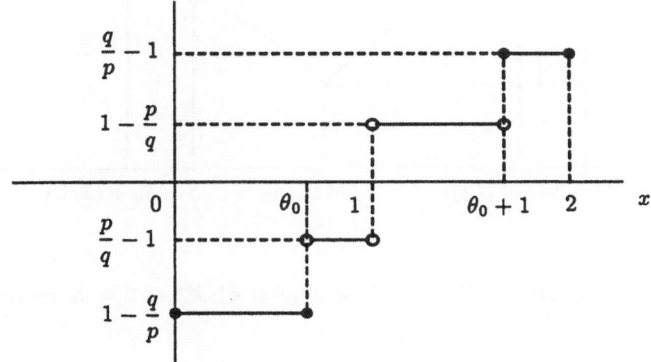

Figure 2.5.2. The values of $\pi_{\theta_0,1}(x) - \pi_{\theta_0,0}(x)$ for all $x \in [0,2]$.

Since for $\theta > \theta_0$,

$$E_\theta \left[\pi_{\theta_0,1}(X) - \pi_{\theta_0,0}(X)\right] = c(\theta - \theta_0),$$

$$E_\theta \left[\{\pi_{\theta_0,1}(X) - \pi_{\theta_0,0}(X)\}^2\right] = c\left\{1 + \frac{(p-q)^2}{pq}(\theta - \theta_0)\right\},$$

we obtain for $\theta > \theta_0$

$$V_\theta \left(\hat{\theta}_1^*(X)\right) = \frac{1}{c} + (\theta - \theta_0)\{1 - (\theta - \theta_0)\},$$

where $V_\theta(\cdot)$ designates the variance at θ. In a similar way to the case when $\theta > \theta_0$, we have for $\theta < \theta_0$

$$V_\theta\left(\hat{\theta}_1^*(X)\right) = \frac{1}{c} + (\theta_0 - \theta)\{1 - (\theta_0 - \theta)\}.$$

Hence the variance of the LMVU estimator $\hat{\theta}_1^*(X)$ is given by

(2.5.8) $$V_\theta\left(\hat{\theta}_1^*(X)\right) = \frac{1}{c} + |\theta - \theta_0|(1 - |\theta - \theta_0|).$$

Note that

$$V_\theta\left(\hat{\theta}_1^*(X)\right) \geq \frac{1}{c} = V_{\theta_0}\left(\hat{\theta}_1^*(X)\right)$$

for all $\theta \in [0, 1]$.

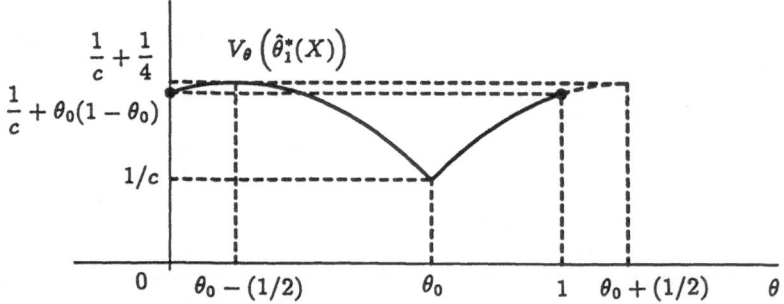

Figure 2.5.3. The variance of the LMVU estimator $\hat{\theta}_1^*(X)$ at $\theta = \theta_0$ given by (2.5.8) when $\theta_0 > 1/2$.

Case II : $n \geq 2$. We shall construct the LMVU estimator at $\theta = \theta_0$ given by the form (2.5.2) in the case when $n \geq 2$. From (2.5.3) we have

(2.5.9) $$\int_0^2 \cdots \int_0^2 \left\{ \prod_{i=1}^n \frac{f(x_i, \eta)}{f(x_i, \theta_0)} \right\} \prod_{i=1}^n f(x_i, \theta) dx_1 \cdots dx_n$$
$$= \{K_{\theta_0}(\theta, \eta)\}^n = K_{\theta_0}^n(\theta, \eta) \quad \text{(say)}.$$

By (2.5.6) and (2.5.7) we obtain for $\theta > \theta_0$

(2.5.10) $$K_{\theta_0}^n(\theta, \eta) = \begin{cases} 1 & \text{for } \eta \leq \theta_0 ; \\ \{1 + c(\eta - \theta_0)\}^n & \text{for } \theta_0 < \eta < \theta ; \\ \{1 + c(\theta - \theta_0)\}^n & \text{for } \theta \leq \eta, \end{cases}$$

and for $\theta < \theta_0$

$$(2.5.11) \qquad K_{\theta_0}^n(\theta, \eta) = \begin{cases} 1 & \text{for} \quad \theta_0 \leq \eta \; ; \\ \{1 + c(\theta_0 - \eta)\}^n & \text{for} \quad \theta < \eta < \theta_0 \; ; \\ \{1 + c(\theta_0 - \theta)\}^n & \text{for} \quad \eta \leq \theta, \end{cases}$$

where $c = \dfrac{1}{pq} - 4$. Note that for $\theta = \theta_0$, $K_{\theta_0}^n(\theta_0, \eta) = 1$. We take a signed measure G_n^* over $[0, 1]$ satisfying

$$(2.5.12) \quad \frac{dG_n^*}{d\eta} = \left(1 - \frac{1}{n}\right) \frac{1}{(1 + c|\eta - \theta_0|)^n} \operatorname{sgn}(\eta - \theta_0) = g_n(\eta) \quad \text{(say)} \quad \text{for} \quad 0 < \eta < \theta_0,$$
$$\theta_0 < \eta < 1;$$

$$G_n^*(\{0\}) = -\frac{1}{nc(1 + c\theta_0)^{n-1}};$$
$$G_n^*(\{1\}) = \frac{1}{nc\{1 + c(1 - \theta_0)\}^{n-1}};$$
$$G_n^*(\{\theta_0\}) = \theta_0.$$

Note that G_1^* in the previous case is consistent with the G_n^* in the case when $n = 1$. Then the estimator

$$(2.5.13) \qquad \hat{\theta}_n^* := \hat{\theta}_n^*(x_1, \ldots, x_n) = \int_0^1 \prod_{i=1}^n \frac{f(x_i, \eta)}{f(x_i, \theta_0)} dG_n^*(\eta)$$

is unbiased. Indeed, we have by (2.5.9), (2.5.10) and (2.5.11) and for $\theta > \theta_0$

$$E_\theta\left(\hat{\theta}_n^*\right) = \int_0^2 \cdots \int_0^2 \left\{ \int_0^1 \prod_{i=1}^n \frac{f(x_i, \eta)}{f(x_i, \theta_0)} dG_n^*(\eta) \right\} \prod_{i=1}^n f(x_i, \theta) dx_1 \cdots dx_n$$

$$= \int_0^1 K_{\theta_0}^n(\theta, \eta) dG_n^*(\eta)$$

$$= G_n^*(\{0\}) K_{\theta_0}^n(\theta, 0) + G_n^*(\{1\}) K_{\theta_0}^n(\theta, 1) + G_n^*(\{\theta_0\}) K_{\theta_0}^n(\theta, \theta_0)$$

$$+ \int_0^{\theta_0} g_n(\eta) d\eta + \int_{\theta_0}^{\theta} \{1 + c(\eta - \theta_0)\}^n g_n(\eta) d\eta$$

$$+ \{1 + c(\theta - \theta_0)\}^n \int_\theta^1 g_n(\eta) d\eta$$

$$= -\frac{1}{nc(1+c\theta_0)^{n-1}} + \frac{\{1+c(\theta-\theta_0)\}^n}{nc\{1+c(1-\theta_0)\}^{n-1}}$$

$$- \left(1-\frac{1}{n}\right)\int_0^{\theta_0} \frac{1}{\{1+c(\theta_0-\eta)\}^n}d\eta + \left(1-\frac{1}{n}\right)\int_{\theta_0}^{\theta} d\eta$$

$$+ \left(1-\frac{1}{n}\right)\{1+c(\theta-\theta_0)\}^n \int_\theta^1 \frac{1}{\{1+c(\eta-\theta_0)\}^n}d\eta + \theta_0$$

$$= \theta.$$

Similarly we obtain for $\theta < \theta_0$

$$E_\theta\left(\hat\theta_n^*\right) = \theta$$

and for $\theta = \theta_0$

$$E_{\theta_0}\left(\hat\theta_n^*\right) = \theta_0.$$

Hence the estimator $\hat\theta_n^*$ is the LMVU estimator.

Next we shall construct the concrete estimator. Putting

$$\pi_{\theta_0,\eta}^n(\mathbf{x}) = \prod_{i=1}^n \frac{f(x_i,\eta)}{f(x_i,\theta_0)} = \prod_{i=1}^n \pi_{\theta_0,\eta}(x_i)$$

with $\mathbf{x} = (x_1,\ldots,x_n)$, we have from (2.5.13)

(2.5.14) $\displaystyle \hat\theta_n^*(\mathbf{x}) = \int_0^1 \pi_{\theta_0,\eta}^n(\mathbf{x})dG_n^*(\eta)$

$$= \pi_{\theta_0,0}^n(\mathbf{x})G_n^*(\{0\}) + \pi_{\theta_0,1}^n(\mathbf{x})G_n^*(\{1\}) + \theta_0$$

$$+ \int_0^{\theta_0} \pi_{\theta_0,\eta}^n(\mathbf{x})g_n(\eta)d\eta + \int_{\theta_0}^1 \pi_{\theta_0,\eta}^n(\mathbf{x})g_n(\eta)d\eta.$$

For $\eta = 0, 1$ we have

$$\pi_{\theta_0,0}^n(\mathbf{x}) = \prod_{i=1}^n \pi_{\theta_0,0}(x_i) = \left(\frac{q}{p}\right)^a \left(\frac{p}{q}\right)^b = \left(\frac{q}{p}\right)^{a-b}$$

and

$$\pi_{\theta_0,1}^n(\mathbf{x}) = \prod_{i=1}^n \pi_{\theta_0,1}(x_i) = \left(\frac{p}{q}\right)^{a'} \left(\frac{q}{p}\right)^{b'} = \left(\frac{p}{q}\right)^{a'-b'}$$

with $a = \#\{x_i\,|\,0 < x_i < \theta_0\}$, $b = \#\{x_i\,|\,1 < x_i < \theta_0+1\}$, $a' = \#\{x_i\,|\,\theta_0 < x_i < 1\}$, $b' = \#\{x_i\,|\,\theta_0+1 < x_i < 2\}$, where $\#\{\ \ \}$ designates the number of the elements of the set $\{\ \ \}$. Note that $0 \le a+b \le n$ and $0 \le a'+b' \le n$. If $0 < \eta < \theta_0$, then

$$\pi_{\theta_0,\eta}(\mathbf{x}) = \left(\frac{q}{p}\right)^s \left(\frac{p}{q}\right)^t = \left(\frac{q}{p}\right)^{s-t},$$

where $s = \#\{x_i \mid \eta < x_i < \theta_0\}$ and $t = \#\{x_i \mid \eta+1 < x_i < \theta_0 + 1\}$. If $\theta_0 < \eta < 1$, then

$$\pi_{\theta_0,\eta}(\mathbf{x}) = \left(\frac{p}{q}\right)^{s'} \left(\frac{q}{p}\right)^{t'} = \left(\frac{p}{q}\right)^{s'-t'},$$

where $s' = \#\{x_i \mid \theta_0 < x_i < \eta\}$ and $t' = \#\{x_i \mid \theta_0 + 1 < x_i < \eta + 1\}$. Note that $0 \leq s + t \leq a + b$ and $0 \leq s' + t' \leq a' + b'$. Hence we have from (2.5.12), (2.5.13) and (2.5.14)

$$(2.5.15) \qquad \hat{\theta}_n^*(\mathbf{x}) = \left(\frac{q}{p}\right)^{a-b} G_n^*(\{0\}) + \left(\frac{p}{q}\right)^{a'-b'} G_n^*(\{1\}) + \theta_0$$

$$+ \int_0^{\theta_0} \left(\frac{q}{p}\right)^{s-t} g_n(\eta)d\eta + \int_{\theta_0}^1 \left(\frac{p}{q}\right)^{s'-t'} g_n(\eta)d\eta$$

$$= -\frac{(q/p)^{a-b}}{nc(1+c\theta_0)^{n-1}} + \frac{(p/q)^{a'-b'}}{nc\{1+c(1-\theta_0)\}^{n-1}} + \theta_0$$

$$+ \int_0^{\theta_0} \left(\frac{q}{p}\right)^{s-t} g_n(\eta)d\eta + \int_{\theta_0}^1 \left(\frac{p}{q}\right)^{s'-t'} g_n(\eta)d\eta,$$

where

$$g_n(\eta) = \left(1 - \frac{1}{n}\right) \frac{1}{(1+c|\eta-\theta_0|)^n} \mathrm{sgn}(\eta - \theta_0) \qquad (0 < \eta < \theta_0, \ \theta_0 < \eta < 1).$$

For $n \geq 2$, it may be difficult to calculate directly the variance of the LMVU estimator $\hat{\theta}_n^*$ given by (2.5.15), but it may be possible to do it by (2.5.13). Putting

$$K_{\theta_0}(\theta, \eta, \eta') = \int_0^2 \pi_{\theta,\eta,\eta'}(x) f(x,\theta) dx$$

with $\pi_{\theta_0,\eta,\eta'}(x) = \pi_{\theta_0,\eta}(x)\pi_{\theta_0,\eta'}(x)$, we have

$$(2.5.16) \ \ E_\theta\left(\hat{\theta}_n^{*2}\right) = \int_0^2 \cdots \int_0^2 \left\{ \int_0^1 \int_0^1 \prod_{i=1}^n \frac{f(x_i,\eta)}{f(x_i,\theta_0)} \prod_{i=1}^n \frac{f(x_i,\eta')}{f(x_i,\theta_0)} dG_n^*(\eta)dG_n^*(\eta') \right\}$$

$$\cdot \prod_{i=1}^n f(x_i,\theta) \prod_{i=1}^n dx_i$$

$$= \int_0^1 \int_0^1 \left\{ \int_0^2 \cdots \int_0^2 \prod_{i=1}^n \frac{f(x_i,\eta)f(x_i,\eta')}{(f(x_i,\theta_0))^2} f(x_i,\theta) \prod_{i=1}^n dx_i \right\}$$

$$\cdot dG_n^*(\eta)dG_n^*(\eta')$$

$$= \int_0^1 \int_0^1 \left\{ \int_0^2 \frac{f(x,\eta)f(x,\eta')}{(f(x,\theta_0))^2} f(x,\theta) dx \right\}^n dG_n^*(\eta)dG_n^*(\eta')$$

$$= \int_0^1 \int_0^1 K_{\theta_0}(\theta, \eta, \eta') dG_n^*(\eta)dG_n^*(\eta').$$

Here we shall obtain the variance of the estimator $\hat{\theta}_n^*$ at $\theta = \theta_0$. By (2.5.4) and (2.5.5) we have for $\eta' < \eta \le \theta_0$

$$(2.5.17) \qquad \pi_{\theta_0,\eta,\eta'}(x) = \begin{cases} 1 & \text{for} \quad 0 \le x \le \eta' \ ; \\ q/p & \text{for} \quad \eta' < x \le \eta \ ; \\ (q/p)^2 & \text{for} \quad \eta < x \le \theta_0 \ ; \\ 1 & \text{for} \quad \theta_0 < x < \eta' + 1 \ ; \\ p/q & \text{for} \quad \eta' + 1 \le x < \eta + 1 \ ; \\ (p/q)^2 & \text{for} \quad \eta + 1 \le x < \theta_0 + 1 \ ; \\ 1 & \text{for} \quad \theta_0 + 1 \le x \le 2. \end{cases}$$

Since $\pi_{\theta_0,\eta,\eta'}(x) = \pi_{\theta_0,\eta',\eta}(x)$, it is easy from (2.5.17) to obtain $\pi_{\theta_0,\eta,\eta'}(x)$ for $\eta < \eta' \le \theta_0$. From (2.5.17) we have

$$(2.5.18) \qquad K_{\theta_0}(\theta_0,\eta,\eta') = \begin{cases} 1 + c(\theta_0 - \eta) & \text{if} \quad \eta' < \eta \le \theta_0 \ ; \\ 1 + c(\theta_0 - \eta') & \text{if} \quad \eta < \eta' \le \theta_0. \end{cases}$$

By (2.5.4) and (2.5.5) we have for $\eta \le \theta_0 \le \eta'$

$$\pi_{\theta_0,\eta,\eta'}(x) = \begin{cases} 1 & \text{for} \quad 0 \le x \le \eta \ ; \\ q/p & \text{for} \quad \eta < x \le \theta_0 \ ; \\ p/q & \text{for} \quad \theta_0 < x \le \eta' \ ; \\ 1 & \text{for} \quad \eta' < x < \eta + 1 \ ; \\ p/q & \text{for} \quad \eta + 1 \le x < \theta_0 + 1 \ ; \\ q/p & \text{for} \quad \theta_0 + 1 \le x < \eta' + 1 \ ; \\ 1 & \text{for} \quad \eta' + 1 \le x \le 2. \end{cases}$$

Then we obtain for $\eta \le \theta_0 \le \eta'$

$$(2.5.19) \qquad K_{\theta_0}(\theta_0,\eta,\eta') = 1.$$

From (2.5.18) and (2.5.19) we have

$$(2.5.20) \qquad K_{\theta_0}(\theta_0,\eta,\eta') = \begin{cases} 1 + c(\theta_0 - \eta) & \text{if} \quad \eta' < \eta \le \theta_0 \ ; \\ 1 + c(\theta_0 - \eta') & \text{if} \quad \eta < \eta' \le \theta_0 \ ; \\ 1 & \text{if} \quad \eta \le \theta_0 \le \eta'. \end{cases}$$

In a similar way to the above we have

$$(2.5.21) \qquad K_{\theta_0}(\theta_0,\eta,\eta') = \begin{cases} 1 & \text{if} \quad \eta' \le \theta_0 \le \eta \ ; \\ 1 + c(\eta' - \theta_0) & \text{if} \quad \theta_0 \le \eta' < \eta \ ; \\ 1 + c(\eta - \theta_0) & \text{if} \quad \theta_0 \le \eta < \eta'. \end{cases}$$

In order to calculate $E_{\theta_0}\left(\hat{\theta}_n^{*2}\right)$ we obtain from (2.5.16)

$$
\begin{aligned}
(2.5.22) \qquad E_{\theta_0}\left(\hat{\theta}_n^{*2}\right) = {} & G_n^*(\{0\}) \int_0^1 K_{\theta_0}^n(\theta_0, 0, \eta') dG_n^*(\eta') \\
& + G_n^*(\{1\}) \int_0^1 K_{\theta_0}^n(\theta_0, 1, \eta') dG_n^*(\eta') \\
& + G_n^*(\{\theta_0\}) \int_0^1 K_{\theta_0}^n(\theta_0, \theta_0, \eta') dG_n^*(\eta') \\
& + \int_0^1 \int_0^{\theta_0} K_{\theta_0}^n(\theta_0, \eta, \eta') dG_n^*(\eta) dG_n^*(\eta') \\
& + \int_0^1 \int_{\theta_0}^1 K_{\theta_0}^n(\theta_0, \eta, \eta') dG_n^*(\eta) dG_n^*(\eta').
\end{aligned}
$$

By (2.5.20) and (2.5.21) we have

$$
(2.5.23) \qquad K_{\theta_0}(\theta_0, 0, \eta') = \begin{cases} 1 + c(\theta_0 - \eta') & \text{for} \quad \eta' \leq \theta_0 ; \\ 1 & \text{for} \quad \theta_0 \leq \eta', \end{cases}
$$

$$
(2.5.24) \qquad K_{\theta_0}(\theta_0, 1, \eta') = \begin{cases} 1 & \text{for} \quad \eta' \leq \theta_0 ; \\ 1 + c(\eta' - \theta_0) & \text{for} \quad \theta_0 \leq \eta'. \end{cases}
$$

It is also clear that

$$
(2.5.25) \qquad K_{\theta_0}(\theta_0, \theta_0, \eta') = 1 \qquad \text{for all} \quad \eta' \in [0, 1].
$$

By (2.5.12), (2.5.23), (2.5.24) and (2.5.25) we have

$$
(2.5.26) \qquad G_n^*(\{0\}) \int_0^1 K_{\theta_0}^n(\theta_0, 0, \eta') dG_n^*(\eta') = 0 ;
$$

$$
(2.5.27) \qquad G_n^*(\{1\}) \int_0^1 K_{\theta_0}^n(\theta_0, 1, \eta') dG_n^*(\eta') = \frac{1}{cn\{1 + c(1 - \theta_0)\}^{n-1}} ;
$$

$$
(2.5.28) \qquad G_n^*(\{\theta_0\}) \int_0^1 K_{\theta_0}^n(\theta_0, \theta_0, \eta') dG_n^*(\eta') = \theta_0^2.
$$

By (2.5.12) and (2.5.25) we also have for $0 < \eta < \theta_0$

$$
\begin{aligned}
\int_0^1 K_{\theta_0}^n(\theta_0, \eta, \eta') dG_n^*(\eta') = {} & K_{\theta_0}^n(\theta_0, \eta, 0) G_n^*(\{0\}) + K_{\theta_0}^n(\theta_0, \eta, 1) G_n^*(\{1\}) \\
& + K_{\theta_0}^n(\theta_0, \eta, \theta_0) G_n^*(\{\theta_0\}) \\
& + \left(\int_0^\eta + \int_\eta^{\theta_0} + \int_{\theta_0}^1 \right) K_{\theta_0}(\theta_0, \eta, \eta') dG_n^*(\eta') \\
= {} & \eta.
\end{aligned}
$$

Then we have for $n \geq 2$

(2.5.29)
$$\int_0^1 \int_0^{\theta_0} K_{\theta_0}^n(\theta_0, \eta, \eta') dG_n^*(\eta) dG_n^*(\eta')$$

$$= \int_0^{\theta_0} \int_0^1 K_{\theta_0}^n(\theta_0, \eta, \eta') dG_n^*(\eta') dG_n^*(\eta)$$

$$= \int_0^{\theta_0} \eta \, dG_n^*(\eta)$$

$$= \begin{cases} -\dfrac{\theta_0}{2c} + \dfrac{1}{2c^2} \log(1 + c\theta_0) & \text{for} \quad n = 2 \text{ ;} \\[2mm] -\dfrac{\theta_0}{cn} + \dfrac{1}{c^2 n(n-2)} - \dfrac{1}{c^2 n(n-2)(1 + c\theta_0)^{n-2}} & \text{for} \quad n > 2. \end{cases}$$

In a similar way to the above we obtain by (2.5.12) and (2.5.21)

(2.5.30)
$$\int_0^1 \int_{\theta_0}^1 K_{\theta_0}^n(\theta_0, \eta, \eta') dG_n^*(\eta) dG_n^*(\eta')$$

$$= \begin{cases} \dfrac{1}{2c} \left[\theta_0 - \dfrac{1}{1 + c(1 - \theta_0)} + \dfrac{1}{c} \log\{1 + c(1 - \theta_0)\} \right] & \text{for} \quad n = 2 \text{ ;} \\[2mm] \dfrac{1}{cn} \left[\theta_0 - \dfrac{1}{\{1 + c(1 - \theta_0)\}^{n-1}} - \dfrac{1}{c^2(n-2)} \left\{ \dfrac{1}{(1 + c(1 - \theta_0))^{n-2}} - 1 \right\} \right] & \text{for} \quad n > 2. \end{cases}$$

From (2.5.22), (2.5.26) to (2.5.30) we have

$$E_{\theta_0}\left(\hat{\theta}_n^{*2}\right)$$

$$= \begin{cases} \theta_0^2 + \dfrac{1}{2c^2} \log\{(1 + c\theta_0)(1 + c(1 - \theta_0))\} & \text{for} \quad n = 2 \text{ ;} \\[2mm] \theta_0^2 + \dfrac{1}{c^2 n(n-2)} \left\{ 2 - \dfrac{1}{(1 + c\theta_0)^{n-2}} - \dfrac{1}{(1 + c(1 - \theta_0))^{n-2}} \right\} & \text{for} \quad n > 2. \end{cases}$$

Hence for $n \geq 2$ the variance of the LMVU estimator $\hat{\theta}_n^*$ is given by

$$V_{\theta_0}\left(\hat{\theta}_n^*\right) = \begin{cases} \dfrac{1}{2c^2} \log\{(1 + c\theta_0)(1 + c(1 - \theta_0))\} & \text{for} \quad n = 2 \text{ ;} \\[2mm] \dfrac{1}{c^2 n(n-2)} \left\{ 2 - \dfrac{1}{(1 + c\theta_0)^{n-2}} - \dfrac{1}{(1 + c(1 - \theta_0))^{n-2}} \right\} & \text{for} \quad n > 2. \end{cases}$$

Note that for sufficiently large n

$$V_{\theta_0}\left(\hat{\theta}_n^*\right) = \frac{2}{c^2 n^2} + o\left(\frac{1}{n^2}\right),$$

that is, the variance of the LMVU estimator $\hat{\theta}_n^*$ is of the order n^{-2}. The calculation of the variance $V_\theta\left(\hat{\theta}_n^*\right)$ at any θ is more complicate than that of $V_{\theta_0}\left(\hat{\theta}_n^*\right)$.

CHAPTER 3

AMOUNTS OF INFORMATION AND THE MINIMUM VARIANCE UNBIASED ESTIMATION

3.1. FISHER INFORMATION AND THE MINIMUM VARIANCE UNBIASED ESTIMATION

In the previous chapter we introduced the concept of the one-directional distribution and discussed the locally minimum variance unbiased estimation for its family. In this section, from another point of view we shall treat the case when the amount of Fisher information is infinity and show that the locally minimum variance of unbiased estimators is equal to zero. Further we shall give two examples.

Let (X, Y) be a pair of random variables defined on some product space $\mathcal{X} \times \mathcal{Y}$ of $\mathcal{X}$ and $\mathcal{Y}$ with a joint probability density function (j.p.d.f.) $f_\theta(x, y)$ with respect to a σ-finite measure $\mu_{x,y}$, where θ is a real-valued parameter. Then it follows that for almost all (x, y) $[\mu_{x,y}]$

$$f_\theta(x, y) = f_\theta(x|y) f_\theta(y),$$

where $f_\theta(x|y)$ and $f_\theta(y)$ denote a conditional p.d.f. of x given y and a marginal p.d.f. of y, respectively. We assume that for almost all (x, y) $[\mu_{x,y}]$, $f_\theta(x, y)$, $f_\theta(x|y)$ and $f_\theta(y)$ are continuously differentiable in θ. Since

$$\log f_\theta(x, y) = \log f_\theta(x|y) + \log f_\theta(y) \qquad \text{a.e.} \quad \mu_{x,y},$$

it follows that

$$I^{X,Y}(\theta) = E_\theta^Y \left[I^{X|Y}(\theta) \right] + I^Y(\theta)$$

where $I^{X,Y}(\theta)$, $I^{X|Y}(\theta)$ and $I^Y(\theta)$ denote the amounts of Fisher information with respect to $f_\theta(x, y)$, $f_\theta(x|y)$ and $f_\theta(y)$, respectively, e.g.,

$$I^{X,Y}(\theta) = \int \int_{\mathcal{X} \times \mathcal{Y}} \left\{ \frac{\partial \log f_\theta(x, y)}{\partial \theta} \right\}^2 f_\theta(x, y) d\mu_{x,y},$$

and $E_\theta^Y(\,\cdot\,)$ is the expectation w.r.t. $f_\theta(y)$. We denote $J_\theta(Y) = I^{X|Y}(\theta)$. Here we assume the following conditions (A.3.1.1) to (A.3.1.3).

(A.3.1.1) $0 < J_\theta(y) < \infty$ for almost all y $[\mu_y]$ and all θ, and $E_{\theta_0}^Y [J_{\theta_0}(Y)] = \infty$ and $I^{X,Y}(\theta_0) = \infty$ for some θ_0.

(A.3.1.2) There exists a sequence $\{B_n\}$ of measurable sets of the domain of Y such that

$$b_n := E_\theta^Y \left[J_{\theta_0}(Y) \chi_{B_n}(Y) \right] < \infty$$

for all θ, b_n is independent of θ and $b_n \to \infty$ as $n \to \infty$, where $\chi_{B_n}(Y)$ denotes the indicator function of the set B_n.

(A.3.1.3) For almost all y $[\mu_y]$, there exists an estimator $\hat{\theta}_y(X)$ of θ such that

$$E_\theta^{X|y} \left[\hat{\theta}_y(X) \right] = \theta \qquad \text{for all} \quad \theta;$$

$$V_\theta^{X|y} \left(\hat{\theta}_y(X) \right) = \frac{1}{J_\theta(y)} \qquad \text{for all} \quad \theta,$$

where $E_\theta^{X|y}(\,\cdot\,)$ and $V_\theta^{X|y}(\,\cdot\,)$ denote the conditional expectation and the conditional variance of X given y, respectively.

We shall show that the locally minimum variance of unbiased estimators of θ is equal to zero.

Theorem 3.1.1. *Assume that the conditions (A.3.1.1) to (A.3.1.3) hold. Then*

$$\inf_{\hat{\theta}(X,Y)\,:\,\text{unbiased}} V_{\theta_0} \left(\hat{\theta}(X,Y) \right) = 0.$$

Proof. Let an estimator $\hat{\theta}^* = \hat{\theta}^*(X,Y)$ be

$$\hat{\theta}^* = \frac{1}{b_n} J_{\theta_0}(Y) \chi_{B_n}(Y) \left\{ \hat{\theta}_Y(X) - \theta_0 \right\} + \theta_0.$$

By the assumptions (A.3.1.2) and (A.3.1.3) we have

$$E_\theta^{X,Y} \left(\hat{\theta}^* \right) = \frac{1}{b_n} E_\theta^Y \left[J_{\theta_0}(Y) \chi_{B_n}(Y) \left\{ E_\theta^{X|Y} \left(\hat{\theta}_Y(X) \right) - \theta_0 \right\} \right] + \theta_0 = \theta$$

for all θ, hence $\hat{\theta}^*$ is an unbiased estimator of θ, where $E_\theta^{X,Y}(\,\cdot\,)$ denotes the expectation w.r.t. $f_\theta(x,y)$. We also obtain by (A.3.1.2) and (A.3.1.3)

$$
\begin{aligned}
E_{\theta_0}^{X,Y} \left(\hat{\theta}^{*2} \right) &= \frac{1}{b_n^2} E_{\theta_0}^{X,Y} \left[J_{\theta_0}^2(Y) \chi_{B_n}(Y) \left\{ \hat{\theta}_Y(X) - \theta_0 \right\}^2 \right] \\
&\quad + \frac{2\theta_0}{b_n} E_{\theta_0}^{X,Y} \left[J_{\theta_0}(Y) \chi_{B_n}(Y) \left\{ \hat{\theta}_Y(X) - \theta_0 \right\} \right] + \theta_0^2 \\
&= \frac{1}{b_n^2} E_{\theta_0}^Y \left[J_{\theta_0}^2(Y) \chi_{B_n}(Y) E_{\theta_0}^{X|Y} \left(\hat{\theta}_Y(X) - \theta_0 \right)^2 \right] \\
&\quad + \frac{2\theta_0}{b_n} E_{\theta_0}^Y \left[J_{\theta_0}(Y) \chi_{B_n}(Y) \left\{ E_{\theta_0}^{X|Y} \left(\hat{\theta}_Y(X) \right) - \theta_0 \right\} \right] + \theta_0^2 \\
&= \frac{1}{b_n^2} E_{\theta_0}^Y \left[J_{\theta_0}(Y) \chi_{B_n}(Y) \right] + \theta_0^2 \\
&= \frac{1}{b_n} + \theta_0^2.
\end{aligned}
$$

Then it follows that the variance of $\hat{\theta}^*$ at θ_0 is given by

$$V_{\theta_0}^{X,Y}\left(\hat{\theta}^*\right) = \frac{1}{b_n},$$

which tends to zero as $n \to \infty$ because of (A.3.1.2). Thus we complete the proof.

3.2. EXAMPLES ON UNBIASED ESTIMATORS WITH ZERO VARI-ANCE

In this section we shall show some examples on the locally minimum variance unbiased estimator with zero variance.

Example 3.2.1. Let $X_i = \theta Y_i + U_i$ $(i = 1, 2, \ldots, n)$, where (Y_i, U_i) $(i = 1, \ldots, n)$ are independent and identically distributed (i.i.d.) random vectors and, for each i, Y_i and U_i are mutually independent. Let $f(u)$ be a known density function of U_i. Assume that

$$I^U := \int \frac{\{f'(u)\}^2}{f(u)} du < \infty.$$

Putting $Y = (Y_1, \ldots, Y_n)$ we have

$$I^Y(\theta) = 0$$

since the distribution of Y is independent of θ, and

$$E^Y \left[J_\theta(Y) \right] = n E \left(Y_1^2 \right) I^U.$$

Assume that $E\left(Y_1^2\right) = \infty$ and $f(u)$ is a density of the standard normal distribution. For any $K > 0$ and each $i = 1, \ldots, n$, we define

$$Y_{i,K}^* = \begin{cases} Y_i & \text{for } |Y_i| \leq K; \\ 0 & \text{for } |Y_i| > K. \end{cases}$$

Since $E\left(Y_1^2\right) = \infty$, it is easily seen that $E\left(Y_{1,K}^{*2}\right)$ tends to infinity as $K \to \infty$. We also define an estimator

$$\hat{\theta}_K = \frac{\sum_{i=1}^n Y_{i,K}^* X_i}{n E \left(Y_{1,K}^{*2} \right)}.$$

It is clear that $\hat{\theta}_K$ is an unbiased estimator of θ. Since the variance of $\hat{\theta}_K$ at $\theta = 0$ is given by

$$V_0\left(\hat{\theta}_K\right) = E_0\left(\hat{\theta}_K^2\right) = \frac{E\left(\sum_{i=1}^n Y_{i,K}^{*2}\right)}{n^2 \left\{ E\left(Y_{1,K}^{*2}\right)\right\}^2} = \frac{1}{n E \left(Y_{1,K}^{*2}\right)},$$

it follows that

$$\lim_{K \to \infty} V_0\left(\hat{\theta}_K\right) = 0,$$

hence

$$\inf_{\hat{\theta}\,:\,\text{unbiased}} V_0\left(\hat{\theta}\right) = 0.$$

Example 3.2.2. Let Y be an unobservable random variable according to the chi-square distribution with one degree of freedom. Assume that $X_1, \ldots, X_n$ are independently, identically, normally and conditionally distributed random variables with mean θ and variance Y given Y, where $n \geq 5$. Then we have

$$I^{X,Y}(\theta) = nE_\theta^{X_1,Y}\left[\left(\frac{X_1-\theta}{Y}\right)^2\right] = nE_\theta^Y\left[\frac{1}{Y^2}E^{X_1|Y}\left[(X_1-\theta)^2\right]\right] = nE_\theta^Y\left(\frac{1}{Y}\right) = \infty,$$

where $X = (X_1, \ldots, X_n)$. For any $K > 0$ we define an estimator

$$\hat{\theta}_K = \begin{cases} \dfrac{c_K\overline{X}}{\sum_{i=1}^n (X_i - \overline{X})^2} & \text{for} \quad \sum_{i=1}^n (X_i - \overline{X})^2 \geq \dfrac{1}{K}; \\[4mm] 0 & \text{for} \quad \sum_{i=1}^n (X_i - \overline{X})^2 < \dfrac{1}{K}, \end{cases}$$

where c_K is a constant such that $\hat{\theta}_K$ is an unbiased estimator of θ. Then we shall show that

$$\lim_{K\to\infty} V_0\left(\hat{\theta}_K\right) = 0.$$

First we easily see that $\sum_{i=1}^n (X_i - \overline{X})^2$ is expressed as a product of two independent random variables Y and Z, where Z has the chi-square distribution with $n-1$ degrees of freedom. Let $f_1(y)$ and $f_{n-1}(z)$ be the density functions of Y and Z, respectively. By the unbiasedness condition of $\hat{\theta}_K$ we obtain

(3.2.1) $$\int\int_{yz \geq 1/K} \frac{1}{yz} f_1(y)f_{n-1}(z)dydz = \frac{1}{c_K}.$$

We also have, as the variance of $\hat{\theta}_K$ at $\theta = 0$,

(3.2.2) $$\int\int_{yz \geq 1/K} \frac{c_K^2}{nyz^2} f_1(y)f_{n-1}(z)dydz = v_K \qquad \text{(say)}.$$

Putting

$$F(Kz) = \int_{1/Kz}^\infty \frac{1}{y} f_1(y)dy,$$

we have

(3.2.3) $$F(Kz) = c'\int_{1/Kz}^\infty y^{-3/2}e^{-y/2}dy = c\sqrt{Kz},$$

where c and c' are constants. By (3.2.1) and (3.2.3) we obtain

$$1 = c_K\int_0^\infty \frac{F(Kz)}{z} f_{n-1}(z)dz = A\sqrt{K}c_K,$$

which implies $c_K = O\left(K^{-1/2}\right)$, where A is some constant. By (3.2.2) and (3.2.3) we also have

$$v_K = c_K^2 \int_0^\infty \frac{F(Kz)}{nz^2} f_{n-1}(z)dz = B\sqrt{K}c_K^2 = O\left(K^{-1/2}\right),$$

where B is some constant. Then we obtain

$$\lim_{K \to \infty} V_0\left(\hat{\theta}_K\right) = \lim_{K \to \infty} v_K = 0.$$

Hence we have

$$\inf_{\hat{\theta}\,:\,\mathrm{unbiased}} V_0\left(\hat{\theta}\right) = 0.$$

3.3. A DEFINITION OF THE GENERALIZED AMOUNT OF INFORMATION

Amount of information contained in a sample and in a statistic (or an estimator) plays an important role in the theory of statistical inference as was shown in papers and books by Fisher (1925, 1934, 1956), Kullback (1959), Papaioannou and Kempthorne (1971) and others. There are, however, various ways of definitions of amount of information, some are more convenient (such as Fisher information) but more restricted in applications.

In this section we shall discuss a definition of the generalized information amount between two distributions which is always well defined, symmetric and additive for independent samples and information contained in a statistic is always not greater than that in the whole sample and the equality holds if and only if the statistic is sufficient. Hence we can also discuss the asymptotic relative efficiency of a statistic (or an estimator) by the ratio of informations contained in the statistic and in the sample in a systematic and unified way both in regular and non-regular cases.

There are various definitions on the distance between two distributions or amounts of information for random variables. Here we consider the following quantity. Let X be a random variable defined over an abstract sample space $\mathcal{X}$ and P and Q are absolutely continuous with respect to a σ-finite measure μ. We define a generalized amount of information between P and Q as

(3.3.1) $$I_X(P,Q) = -8\log \int \left(\frac{dP}{d\mu}\frac{dQ}{d\mu}\right)^{1/2} d\mu.$$

Here the integral in the above is called affinity between P and Q (see, e.g., Matusita, 1955). The same amount of information as the above is also discussed by LeCam (1990) in relation to the affinity (see also LeCam (1986)). An extension of the amount of information to as Rényi measure and the asymptotic loss of information are studied by Akahira (1994b). The above quantity (3.3.1) is independent of a choice of the measure μ. Indeed, if P and Q are absolutely continuous with respect to a σ-finite measure ν, then they are also so w.r.t. $r := \mu + \nu$, hence

$$\int \left(\frac{dP}{d\mu}\frac{dQ}{d\mu}\right)^{1/2} d\nu = \int \left(\frac{dP}{d\nu}\frac{dQ}{d\nu}\right)^{1/2} d\mu = \int \left(\frac{dP}{dr}\frac{dQ}{dr}\right)^{1/2} dr.$$

Taking $P + Q$ as μ, we can see that $I_X(P, Q)$ is always well defined since P and Q are absolutely continuous w.r.t. μ. If P and Q are not equivalent, then $I_X(P, Q) > 0$. So far as P and Q are not disjoint, that is, for any measurable set A,

$$P(A)Q(A) + (1 - P(A))(1 - Q(A)) > 0,$$

it follows that $I_X(P, Q)$ is finite. When P and Q are disjoint, we define $I_X(P, Q) = \infty$.

If X and Y are independent random variables and have the distributions P_i and Q_i $(i = 1, 2)$, then it is easily seen that

$$I_{X,Y}(P_1 \times Q_1, P_2 \times Q_2) = I_X(P_1, P_2) + I_Y(Q_1, Q_2),$$

where $P_1 \times Q_1$ and $P_2 \times Q_2$ denote product measures of P_i $(i = 1, 2)$ and Q_i $(i = 1, 2)$, respectively.

Let $T = t(X)$ be a statistic. We denote by $P_{X|T}$, $Q_{X|T}$ and P_T, Q_T the conditional distributions of X given T and the distributions of T, which are assumed to be absolutely continuous w.r.t. σ-finite measures μ_1 and μ_2, respectively. Then we obtain the following.

Theorem 3.3.1. *It holds that*

$$I_T(P, Q) \leq I_X(P, Q),$$

where the equality holds if and only if T is pairwise sufficient for P and Q.

Proof. We have

$$\int \left(\frac{dP_X}{d\mu} \cdot \frac{dQ_X}{d\mu} \right)^{1/2} d\mu$$

$$= \int \int \left(\frac{dP_{X|T}}{d\mu_1} \cdot \frac{dP_T}{d\mu_2} \right)^{1/2} \left(\frac{dQ_{X|T}}{d\mu_1} \cdot \frac{dQ_T}{d\mu_2} \right)^{1/2} d\mu_1 d\mu_2$$

$$= \int \left\{ \int \left(\frac{dP_{X|T}}{d\mu_1} \cdot \frac{dQ_{X|T}}{d\mu_1} \right)^{1/2} d\mu_1 \right\} \left(\frac{dP_T}{d\mu_2} \cdot \frac{dQ_T}{d\mu_2} \right)^{1/2} d\mu_2$$

$$\leq \int \left(\frac{dP_T}{d\mu_2} \cdot \frac{dQ_T}{d\mu_2} \right)^{1/2} d\mu_2,$$

hence by (3.3.1)

$$I_T(P, Q) \leq I_X(P, Q).$$

In the above the equality holds if and only if $P_{X|T} = Q_{X|T}$ a.a. T, that is, T is pairwise sufficient for P and Q. Thus we complete the proof.

Next we define a new distribution K_T of T as

$$\frac{dK_T}{d\mu_2} = c \left(\frac{dP_T}{d\mu_2} \cdot \frac{dQ_T}{d\mu_2} \right)^{1/2}, \qquad \int dK_T = 1,$$

where c is some constant. Since

$$\int \left(\frac{dP_X}{d\mu} \cdot \frac{dQ_X}{d\mu} \right)^{1/2} d\mu$$

$$= \left[\int \left\{ \exp \left(-\frac{1}{8} I_{X|T}(P,Q) \right) \right\} \frac{dK_T}{d\mu_2} d\mu_2 \right] \cdot \exp \left\{ -\frac{1}{8} I_T(P,Q) \right\},$$

it follows that

$$(3.3.2) \qquad I_X(P,Q) = -8 \log E_T^* \left[\exp \left\{ -\frac{1}{8} I_{X|T}(P,Q) \right\} \right] + I_T(P,Q),$$

where $I_{X|T}(P,Q)$ is the amount of information between the conditional distributions of X given T, and E^* denotes the expectation w.r.t. the distribution K_T.

The constant 8 in the amount $I_X(P,Q)$ of information is given to have a connection with that of the Fisher information. Indeed, we consider a family $\mathcal{P} := \{ P_\theta \,|\, \theta \in \Theta \}$ of distributions of X, and suppose that for each $\theta \in \Theta$ P_θ is absolutely continuous w.r.t. a σ-finite measure μ. Then we denote the density $dP_\theta/d\mu$ by $f(\cdot, \theta)$, and the amount of information between P_{θ_1} and P_{θ_2} in $\mathcal{P}$ by $I(\theta_1, \theta_2)$. Under suitable regularity conditions on $f(x, \theta)$, we have for sufficiently small $\Delta\theta$

$$I(\theta, \theta + \Delta\theta) = -8 \log \left\{ \int \left(\frac{dP_\theta}{d\mu} \cdot \frac{dP_{\theta+\Delta\theta}}{d\mu} \right)^{1/2} d\mu \right\}$$

$$= -8 \log \left[\int \{ f(x,\theta) f(x, \theta + \Delta\theta) \}^{1/2} d\mu \right]$$

$$= -8 \log \left[\int \left\{ 1 + \frac{f(x, \theta + \Delta\theta) - f(x,\theta)}{f(x,\theta)} \right\}^{1/2} f(x,\theta) d\mu \right]$$

$$= -8 \log \left[1 - \frac{1}{8} \int \left\{ \frac{f(x, \theta + \Delta\theta) - f(x,\theta)}{f(x,\theta)} \right\}^2 f(x,\theta) d\mu \right] + o\left((\Delta\theta)^2 \right)$$

$$= \int \left\{ \frac{f(x, \theta + \Delta\theta) - f(x,\theta)}{f(x,\theta)} \right\}^2 f(x,\theta) d\mu + o\left((\Delta\theta)^2 \right)$$

$$= \int \left[\frac{\{ (\partial/\partial\theta) f(x,\theta) \} \Delta\theta}{f(x,\theta)} \right]^2 f(x,\theta) d\mu + o\left((\Delta\theta)^2 \right)$$

$$= I(\theta) (\Delta\theta)^2 + o\left((\Delta\theta)^2 \right),$$

where $I(\theta)$ denotes the amount of Fisher information.

Suppose that $X_1, \ldots, X_n$ are independent and identically distributed (i.i.d.) random variables with a density function $f(x, \theta)$, where θ is a real-valued parameter. Then, for θ and $\theta + \Delta\theta$, the amount of information for $(X_1, \ldots, X_n)$ is given by $nI(\theta, \theta + \Delta\theta)$.

Suppose furthermore that there exists a consistent estimator $T_n = t(X_1, \ldots, X_n)$ with the order c_n. Then it follows that

$$\lim_{n \to \infty} I_{T_n}(\theta, \theta + c_n^{-1}\Delta) > 0$$

for some nonzero constant Δ, and

(3.3.3) $$\lim_{n \to \infty} nI(\theta, \theta + c_n^{-1}\Delta) > 0$$

which gives the bound for consistency (see Example 3.4.5 and also Section 3.5).

3.4. EXAMPLES ON THE GENERALIZED AMOUNT OF INFORMA- TION

In this section we have several examples on the generalized amount of information and discuss some details on the structure of information.

Example 3.4.1. Suppose that, for two distributions P_j $(j = 1, 2)$, $X_1, \ldots, X_n$ are independently, identically and normally distributed (i.i.n.d.) with mean μ_j and variance σ_j^2. Then the amount of information between two distributions for each X_i is given by

$$
\begin{aligned}
I(1, 2) &= -8 \log \left[\int \frac{1}{\sqrt{2\pi}\sigma_1 \sigma_2} \exp \left\{ -\frac{\sigma_2^2 (x - \mu_1)^2 + \sigma_1^2 (x - \mu_2)^2}{4\sigma_1^2 \sigma_2^2} \right\} dx \right] \\
&= -8 \log \left[\sqrt{\frac{2\sigma_1 \sigma_2}{\sigma_1^2 + \sigma_2^2}} \exp \left\{ -\frac{\mu_1 - \mu_2}{4(\sigma_1^2 + \sigma_2^2)} \right\} \right] \\
&= 4 \log \left(\frac{\sigma_1^2 + \sigma_2^2}{2\sigma_1 \sigma_2} \right) + \frac{2(\mu_1 - \mu_2)^2}{\sigma_1^2 + \sigma_2^2}.
\end{aligned}
$$

Hence it is seen that the amount of information for $(X_1, \ldots, X_n)$ is equal to n-times the above value. We put $T_1 = \overline{X} = \sum_{i=1}^n X_i/n$ and $T_2 = \sum_{i=1}^n (X_i - \overline{X})^2$. Since

$$I_{T_1}(1, 2) = 4 \log \left(\frac{\sigma_1^2 + \sigma_2^2}{2\sigma_1 \sigma_2} \right) + \frac{2n(\mu_1 - \mu_2)^2}{\sigma_1^2 + \sigma_2^2}$$

and

$$I_{T_2}(1, 2) = 4(n - 1) \log \left(\frac{\sigma_1^2 + \sigma_2^2}{2\sigma_1 \sigma_2} \right),$$

it follows that

$$I_{T_1}(1, 2) + I_{T_2}(1, 2) = nI(1, 2).$$

The equality should hold since (T_1, T_2) is a sufficient statistic.

Example 3.4.2. Suppose that, for two distributions P_{θ_j} $(j = 1, 2)$, $X_1, \ldots, X_n$ are i.i.d. random variables according to a t-distribution with 3 degrees of freedom, i.e., with the density

$$f(x - \theta_j) = \frac{c}{\left(1 + (x - \theta_j)^2 \right)^2},$$

where c is some constant. Then the amount of information between two distributions for each X_i is given by

$$I(1,2) = -8\log\left(1 + \frac{(\theta_1 - \theta_2)^2}{4}\right).$$

Example 3.4.3. Suppose that, for two distributions P_{θ_j} $(j = 1, 2)$, $X_1, \ldots, X_n$ are i.i.d. random variables with an exponential density

$$f(x, \theta_j, \xi_j) = \begin{cases} \dfrac{1}{\theta_j}\exp\left(-\dfrac{x - \xi_j}{\theta_j}\right) & \text{for } x \geq \xi_j, \\ 0 & \text{for } x < \xi_j, \end{cases}$$

where $\theta_j > 0$ and $-\infty < \xi_j < \infty$ $(j = 1, 2)$. If $\xi_1 < \xi_2$, then the amount of information between two distributions for each X_i is given by

$$
\begin{aligned}
I(1,2) &= -8\log\left[\int_{\xi_2}^\infty \frac{1}{(\theta_1\theta_2)^{1/2}}\exp\left\{-\frac{1}{2}\left(\frac{x - \xi_1}{\theta_1} + \frac{x - \xi_2}{\theta_2}\right)\right\}dx\right] \\
&= -8\log\left\{\frac{2\sqrt{\theta_1\theta_2}}{\theta_1 + \theta_2}\cdot\exp\left(-\frac{\xi_2 - \xi_1}{2\theta_1}\right)\right\} \\
&= 4\log\frac{(\theta_1 + \theta_2)^2}{4\theta_1\theta_2} + \frac{4(\xi_2 - \xi_1)}{\theta_1}.
\end{aligned}
$$

We put $T_1 = \min_{1 \leq i \leq n} X_i$ and $T_2 = \overline{X} - \min_{1 \leq i \leq n} X_i$. Since T_1 is distributed according to an exponential distribution and it can be written that

$$nT_2 = \sum_{i=1}^n X_{(i)} - nX_{(1)} = \sum_{i=1}^{n-1}(n - i)\left(X_{(i+1)} - X_{(i)}\right)$$

which is equal to the sum of $n - 1$ i.i.d. exponential random variables, it follows that

$$(3.4.1) \qquad I_{T_1}(1,2) = 4\log\frac{(\theta_1 + \theta_2)^2}{4\theta_1\theta_2} + 4n\frac{\xi_2 - \xi_1}{\theta_1},$$

$$(3.4.2) \qquad I_{T_2}(1,2) = 4(n - 1)\log\frac{(\theta_1 + \theta_2)^2}{4\theta_1\theta_2},$$

where $X_{(1)} \leq X_{(2)} \leq \cdots \leq X_{(n)}$. Then we have

$$(3.4.3) \qquad I_{T_1, T_2}(1,2) = I_{T_1}(1,2) + I_{T_2}(1,2) = nI(1,2),$$

which shows that the pair (T_1, T_2) is sufficient. And also (3.4.1) and (3.4.2) show that T_1 is sufficient when θ is known, i.e., $\theta = \theta_1 = \theta_2$, and T_1 has only $1/n$ of the total information when ξ is known, i.e., $\xi = \xi_1 = \xi_2$. This fact can be interpreted as showing that T_2 is asymptotically sufficient when ξ is known.

Example 3.4.4. Suppose that, for two distributions P_j $(j = 1, 2)$, $X_1, \ldots, X_n$ are i.i.d. random variables according to a uniform distribution on the interval $[\theta_j - \tau_j/2,$

$\theta_j + \tau_j/2\big]$. If $\tau_1 < \tau_2$, then the amount of information between two distributions for each X_i is given as

$$
I(1,2) = \begin{cases}
\infty & \text{for}\quad \theta_1 + \dfrac{\tau_1}{2} \le \theta_2 - \dfrac{\tau_2}{2}\ \text{or}\ \theta_1 - \dfrac{\tau_1}{2} \ge \theta_2 + \dfrac{\tau_2}{2}, \\[2mm]
-4\log\left\{\dfrac{1}{\tau_1\tau_2}\left(\dfrac{\tau_1+\tau_2}{2} + \theta_1 - \theta_2\right)^2\right\} \\[1mm]
\quad\text{for}\quad \theta_1 - \dfrac{\tau_1}{2} < \theta_2 - \dfrac{\tau_2}{2} < \theta_1 + \dfrac{\tau_1}{2} < \theta_2 + \dfrac{\tau_2}{2}, \\[2mm]
-4\log\left\{\dfrac{1}{\tau_1\tau_2}\left(\dfrac{\tau_1+\tau_2}{2} + \theta_2 - \theta_1\right)^2\right\} \\[1mm]
\quad\text{for}\quad \theta_2 - \dfrac{\tau_2}{2} < \theta_1 - \dfrac{\tau_1}{2} < \theta_2 + \dfrac{\tau_2}{2} < \theta_1 + \dfrac{\tau_1}{2}, \\[2mm]
-4\log\dfrac{\tau_1}{\tau_2} & \text{for}\quad \theta_2 - \dfrac{\tau_2}{2} < \theta_1 - \dfrac{\tau_1}{2} < \theta_1 + \dfrac{\tau_1}{2} < \theta_2 + \dfrac{\tau_2}{2}.
\end{cases}
$$

We put $T_1 = \max_{1\le i\le n} X_i - \min_{1\le i\le n} X_i$ and $T_2 = (\max_{1\le i\le n} X_i + \min_{1\le i\le n} X_i)/2$. Then the joint density of (T_1, T_2) is given by

$$
f(t_1, t_2) = \begin{cases}
\dfrac{n(n-1)}{\tau^n} t_1^{n-2} & \text{for}\quad t_2 - \dfrac{t_1}{2} > \theta - \dfrac{\tau}{2}\ \text{or}\ t_2 + \dfrac{t_1}{2} < \theta + \dfrac{\tau}{2}, \\[2mm]
0 & \text{otherwise.}
\end{cases}
$$

Hence the density of T_1 is obtained by

$$
f(t_1) = \begin{cases}
n(n-1)t_1^{n-2}(\tau - t_1)/\tau^n & \text{for}\quad 0 < t_1 < \tau, \\
0 & \text{otherwise,}
\end{cases}
$$

and the amount of information between two distributions is also given by

$$
I_{T_1}(1,2) = 4n\log\dfrac{\tau_2}{\tau_1} \qquad \text{for}\quad \tau_1 < \tau_2.
$$

If $\tau_1 = \tau_2 = \tau$, it is easily seen that $I_{T_1}(1,2) = 0$. However, it should be noted that T_2 alone can not be sufficient even when τ is known, i.e., $\tau = \tau_1 = \tau_2$ since T_1 and T_2 are not independent. Then the amount of information for $(X_1, \ldots, X_n)$ is also given by

$$
(3.4.4) \qquad nI(1,2) = \begin{cases}
-8n\log\left(1 - \dfrac{|\theta_1 - \theta_2|}{\tau}\right) & \text{for}\quad |\theta_1 - \theta_2| < \tau, \\[2mm]
\infty & \text{for}\quad |\theta_1 - \theta_2| \ge \tau.
\end{cases}
$$

Since the conditional distribution of T_2 given T_1 is a uniform distribution on the interval $[\theta - (\tau - T_1)/2, \theta + (\tau - T_1)/2]$, it follows that the conditional information amount is given by

$$
I_{T_2|T_1}(1,2) = \begin{cases}
-8\log\left(1 - \dfrac{|\theta_1 - \theta_2|}{\tau - T_1}\right) & \text{for}\quad |\theta_1 - \theta_2| < \tau - T_1, \\[2mm]
\infty & \text{for}\quad |\theta_1 - \theta_2| \ge \tau - T_1.
\end{cases}
$$

Then we have

$$E\left[\exp\left\{-\frac{1}{8}I_{T_2|T_1}(1,2)\right\}\right] = E\left[\left(\frac{\tau - T_1 - |\theta_1 - \theta_2|}{\tau - T_1}\right)^+\right]$$

$$= \int_0^{\tau - |\theta_1 - \theta_2|} n(n-1)(\tau - t_1 - \theta_1 - \theta_2)t_1^{n-2}/\tau^n dt_1$$

$$= \left(1 - \frac{|\theta_1 - \theta_2|}{\tau}\right)^n,$$

where $(\ )^+$ denote a positive part of $(\ )$. Hence

$$-8\log E_{T_1}\left[\exp\left\{-\frac{1}{8}I_{T_2|T_1}(1,2)\right\}\right] = -8n\log\left(1 - \frac{|\theta_1 - \theta_2|}{\tau}\right)$$

for $|\theta_1 - \theta_2| < \tau$, which is equal to (3.4.4) as is expected from (3.3.2) since $I_{T_1}(1,2) = 0$. We also obtain, as the density of T_2,

$$f_{T_2}(t_2) = \int_0^{\tau - 2|t_2 - \theta|} f(t_1, t_2)dt_1 = \frac{n}{\tau^n}(\tau - 2|t_2 - \theta|)^{n-1}$$

for $|t_2 - \theta| < \tau/2$. Hence the amount of information for T_2 is given by

$$I_{T_2}(1,2) = -8\log\left[\int_{\theta_1 - \tau/2}^{\theta_2 + \tau/2} \frac{n}{\tau^n}(\tau - 2|t_2 - \theta_1|)^{(n-1)/2}(\tau - 2|t_2 - \theta_2|)^{(n-1)/2}dt_2\right]$$

for $0 < |\theta_2 - \theta_1| < \tau$. The explicit form of the above integral is complicated, so we consider the case when n is large and θ_1 and θ_2 are sufficiently close. Putting $\Delta/n = |\theta_2 - \theta_1|$, we have for sufficiently large n

$$(3.4.5) \qquad I_{T_2}(1,2) \approx -8\log\int_{-\infty}^{\infty}\frac{1}{\tau}\exp\left\{-\frac{1}{\tau}(|t_2| + |t_2 - \Delta|)\right\}dt_2$$

$$= 8\left\{\frac{\Delta}{\tau} - \log\left(1 + \frac{\Delta}{\tau}\right)\right\}.$$

On the other hand we obtain from (3.4.4)

$$nI(1,2) \approx -8n\log\left(1 - \frac{\Delta}{n\tau}\right) \sim \frac{8\Delta}{\tau}$$

for sufficiently large n. Therefore the loss of information of T_2 is asymptotically equal to the value $8\log(1 + (\Delta/\tau))$.

Example 3.4.5. Suppose that $X_1, \ldots, X_n$ are i.i.d. random variables with a triangular density function

$$f(x - \theta) = \begin{cases} 1 - |x - \theta| & \text{for} \quad |x - \theta| < 1, \\ 0 & \text{for} \quad |x - \theta| \geq 1. \end{cases}$$

For a small positive number $\Delta\theta$, we have

$$\int_{-1+\Delta\theta}^{1} \{f(x)f(x-\Delta\theta)\}^{1/2}\, dx$$

$$= \int_{-1+\Delta\theta}^{0} \{(1+x)(1+x-\Delta\theta)\}^{1/2}\, dx + \int_{0}^{\Delta\theta} \{(1-x)(1+x-\Delta\theta)\}^{1/2}\, dx$$

$$+ \int_{\Delta\theta}^{1} \{(1-x)(1-x+\Delta\theta)\}^{1/2}\, dx$$

$$= 2\int_{0}^{1-\Delta\theta} \sqrt{y^2 + y\Delta\theta}\, dy + \int_{0}^{\Delta\theta} \{(1-x)(1+x-\Delta\theta)\}^{1/2}\, dx$$

$$= 2\int_{\Delta\theta/2}^{1-(\Delta\theta/2)} \sqrt{u^2 - \frac{(\Delta\theta)^2}{4}}\, du + \int_{-\Delta\theta/2}^{\Delta\theta/2} \sqrt{\left(1-\frac{\Delta\theta}{2}\right)^2 - u^2}\, du$$

$$= \sqrt{\left(1-\frac{\Delta\theta}{2}\right)^2 - \frac{(\Delta\theta)^2}{4}} + \frac{(\Delta\theta)^2}{4}\log\frac{\Delta\theta}{4}$$

$$- \frac{(\Delta\theta)^2}{4}\log\frac{1}{2}\left|1 - \frac{\Delta\theta}{2}\sqrt{\left(1-\frac{\Delta\theta}{2}\right)^2 - \frac{(\Delta\theta)^2}{4}}\right|$$

$$+ \left(1-\frac{\Delta\theta}{2}\right)^2 \sin^{-1}\frac{\Delta\theta/2}{1-(\Delta\theta/2)}$$

$$= 1 + \frac{(\Delta\theta)^2}{4}\log\frac{\Delta\theta}{4} - \frac{3}{8}(\Delta\theta)^2 + o\left((\Delta\theta)^2\right).$$

In a similar way to the case $\Delta\theta > 0$, we obtain for $\Delta\theta < 0$,

$$\int_{-1}^{1+\Delta\theta} \{f(x)f(x-\Delta\theta)\}^{1/2}\, dx = 1 + \frac{(\Delta\theta)^2}{4}\log\frac{|\Delta\theta|}{4} - \frac{3}{8}(\Delta\theta)^2 + o\left((\Delta\theta)^2\right).$$

Hence we have for small $\Delta\theta$ $(\neq 0)$

$$I(\theta, \theta + \Delta\theta) = -8\log\int \{f(x)f(x-\Delta\theta)\}^{1/2}\, dx$$

$$= -8\log\left(1 + \frac{(\Delta\theta)^2}{4}\log\frac{|\Delta\theta|}{4} - \frac{3}{8}(\Delta\theta)^2 + o\left((\Delta\theta)^2\right)\right)$$

$$= -2(\Delta\theta)^2\log\frac{|\Delta\theta|}{4} + 3(\Delta\theta)^2 + o\left((\Delta\theta)^2\right).$$

In order that $nI(\theta, \theta + \Delta\theta) = O(1)$, i.e. $-n(\Delta\theta)^2\log|\Delta\theta| = O(1)$, we obtain $\Delta\theta = O\left((n\log n)^{-1/2}\right)$. Letting $c_n = (n\log n)^{1/2}$, we see that (3.3.3) holds, i.e. $\underline{\lim}_{n\to\infty} nI(\theta, \theta + c_n^{-1}\Delta) > 0$ for some nonzero constant Δ. Hence it follows that the bound for consistency is equal to $(n\log n)^{1/2}$.

Example 3.4.6. Suppose that $X_1, \ldots, X_n$ are i.i.d. random variables according to a distribution P_θ with the truncated normal density

$$f(x - \theta) = \begin{cases} ce^{-(x-\theta)^2/2} & \text{for } |x - \theta| < 1, \\ 0 & \text{for } |x - \theta| \geq 1, \end{cases}$$

where c is some constant. For $0 < \theta < 2$, the affinity $\rho(0, \theta)$ between P_0 and P_θ is given by

$$\rho(0, \theta) = \int_{-\infty}^{\infty} \left(\frac{dP_0}{dx} \cdot \frac{dP_\theta}{dx} \right)^{1/2} dx$$

$$= \int_{\theta-1}^{1} c \exp\left[-\frac{1}{4} \left\{ x^2 + (x - \theta)^2 \right\} \right] dx$$

$$= e^{-\theta^2/8} P_0 \left\{ |X_1| < 1 - \frac{\theta}{2} \right\}.$$

For small $\theta > 0$, the amount of information between P_0 and P_θ for X_1 is given by

$$(3.4.6) \qquad I_{X_1}(0, \theta) = \theta^2 - 8 \log P_0 \left\{ |X_1| < 1 - \frac{\theta}{2} \right\}$$

$$= \theta^2 - 8 \log \left[1 - 2\sqrt{2\pi} c \left\{ \Phi(1) - \Phi\left(1 - \frac{\theta}{2}\right) \right\} \right]$$

$$= \theta^2 - 8 \log \left(1 - k\theta - \frac{1}{4} k\theta^2 + o(\theta^2) \right)$$

$$= 8k\theta + \{1 + 2k(2k + 1)\} \theta^2 + o\left(\theta^2\right),$$

where $\Phi(x)$ is the standard normal distribution function. Since $\overline{X} := \sum_{i=1}^{n} X_i/n$ is asymptotically and normally distributed with mean θ and variance $(1 - 2k)/n$, it follows that for large n the amount of information between two distributions for $\overline{X}$ is given by

$$(3.4.7) \qquad I_{\overline{X}}(0, \theta) \approx -8 \log \int_{-\infty}^{\infty} \frac{1}{\sqrt{2\pi}} \sqrt{\frac{n}{1 - 2k}} \exp\left[-\frac{n}{4(1 - 2k)} \left\{ x^2 + (x - \theta)^2 \right\} \right] dx$$

$$= \frac{n\theta^2}{1 - 2k} = O\left(n\theta^2\right),$$

where $k = ce^{-1/2}$. We put $T_1 = n\left(X_{(1)} + 1 - \theta\right)$ and $T_2 = n\left(X_{(n)} - 1 - \theta\right)$, where $X_{(1)} = \min_{1 \leq i \leq n} X_i$ and $X_{(n)} = \max_{1 \leq i \leq n} X_i$. Since the asymptotic densities of T_1 and T_2 are given by

$$g_1(t) = \begin{cases} ke^{-kt} & \text{for } t > 0; \\ 0 & \text{for } t \leq 0, \end{cases}$$

and

$$g_2(t) = \begin{cases} ke^{kt} & \text{for } t < 0; \\ 0 & \text{for } t \geq 0, \end{cases}$$

respectively, it follows that for large n and $\theta > 0$, the amounts of information between two distributions for $X_{(1)}$ and $X_{(n)}$ are given by

$$(3.4.8) \qquad I_{T_1}\left(0, \frac{\theta}{n}\right) \approx -8\log\left(\int_{\theta}^{\infty} k e^{-kt+(k\theta/2)} dt\right) = 4k\theta$$

and

$$(3.4.9) \qquad I_{T_2}\left(0, \frac{\theta}{n}\right) \approx -8\log\left(\int_{-\infty}^{0} k e^{kt-(k\theta/2)} dt\right) = 4k\theta.$$

It is noted that $\overline{X}$ is asymptotically independent of $X_{(1)}$ and $X_{(n)}$, and $\left(X_{(1)}, X_{(n)}\right)$ is asymptotically sufficient. It follows from (3.4.6) to (3.4.9) that for large n and small $\theta > 0$ the amount of information between two distributions for $(X_1, \ldots, X_n)$ has the following relationship :

$$\begin{aligned} I_{X_1,\ldots,X_n}(0,\theta) &= n I_{X_1}(0,\theta) \\ &= I_{X_{(1)}}(0,\theta) + I_{X_{(n)}}(0,\theta) + O\left(n\theta^2\right). \end{aligned}$$

In order to consider the above up to the order $o\left(n\theta^2\right)$, it is necessary to discuss the loss of information associated with the statistics up to the higher order and its relation to the sufficiency (see Akahira 1991b, 1994b).

Example 3.4.7. Suppose that for $n = 2m + 1$, $X_1, \ldots, X_n$ are independent and identically distributed random variables according to a density $f(x - \theta)$ w.r.t. the Lebesgue measure. Let T be the median of $X_1, \ldots, X_n$, symbolically $T = \text{med} X_i$. We shall obtain the amount of information between two distributions for T and the asymptotic relative efficiency of T w.r.t. $(X_1, \ldots, X_n)$. First the density $g(t - \theta)$ of T is given by

$$g(t - \theta) = \frac{(2m + 1)!}{m!m!} \left\{F(t - \theta)\right\}^m \left\{1 - F(t - \theta)\right\}^m f(t - \theta),$$

where $F(x - \theta)$ denotes the distribution function of X_1. Then it follows that for $\theta_1 = \theta + \Delta$ and $\theta_2 = \theta - \Delta$, the affinity $\rho(\theta_1, \theta_2)$ between two distributions with densities $g(t - \theta_j)$ $(j = 1, 2)$ is given by

$$\rho(\theta_1, \theta_2) = \frac{(2m + 1)!}{m!m!} \int_{-\infty}^{\infty} \left\{F(t - \Delta)F(t + \Delta)\left(1 - F(t - \Delta)\right)\left(1 - F(t + \Delta)\right)\right\}^{m/2}$$

$$\cdot \left\{f(t - \Delta)f(t + \Delta)\right\}^{1/2} dt.$$

Putting

$$H(t) = F(t - \Delta)F(t + \Delta)\left(1 - F(t - \Delta)\right)\left(1 - F(t + \Delta)\right),$$

we have

$$\frac{d}{dt}\log H(t) = \frac{f(t - \Delta)}{F(t - \Delta)} + \frac{f(t + \Delta)}{F(t + \Delta)} - \frac{f(t - \Delta)}{1 - F(t - \Delta)} - \frac{f(t + \Delta)}{1 - F(t + \Delta)}.$$

If $f(-t) = f(t)$ for all real number t and $f(t)/(1 - F(t))$ is a monotone increasing function of t, it follows that four terms of the right-hand side of the above are monotone decreasing functions of t, which implies that $(d/dt)\log H(t)$ is a monotone decreasing function of t. Since $f(-\Delta) = f(\Delta)$ and $F(\Delta) = 1 - F(-\Delta)$, it follows that

$$\frac{d}{dt}\log H(0) = \frac{f(-\Delta)}{F(-\Delta)} + \frac{f(\Delta)}{F(\Delta)} - \frac{f(-\Delta)}{1 - F(-\Delta)} - \frac{f(\Delta)}{1 - F(\Delta)} = 0,$$

implying that $\log H(t)$, i.e., $H(t)$ takes a maximum value at $t = 0$. We also have

$$\frac{d^2}{dt^2}\log H(0) = \frac{f'(-\Delta)}{F(-\Delta)} - \frac{f^2(-\Delta)}{F^2(-\Delta)} + \frac{f'(\Delta)}{F(\Delta)} - \frac{f^2(\Delta)}{F^2(\Delta)}$$
$$- \frac{f'(-\Delta)}{1 - F(-\Delta)} - \frac{f^2(-\Delta)}{(1 - F(-\Delta))^2} - \frac{f'(\Delta)}{1 - F(\Delta)} - \frac{f^2(\Delta)}{(1 - F(\Delta))^2}$$
$$= \frac{2f'(\Delta)(1 - 2F(\Delta))}{F(\Delta)(1 - F(\Delta))} - \frac{2f^2(\Delta)\left\{F^2(\Delta) + (1 - F(\Delta))^2\right\}}{F^2(\Delta)(1 - F(\Delta))^2}$$
$$= -\sigma^2 \qquad \text{(say)}.$$

Then we obtain

$$\log H(t) = \log H(0) - \frac{\sigma^2}{2}t^2 + \cdots.$$

Hence we have for sufficiently large m

$$\rho(\theta_1, \theta_2) \approx \frac{(2m + 1)!}{m!m!} H^{m/2}(0) \int_{-\infty}^{\infty} e^{-\sigma^2 mt^2/4} \{f(t - \Delta)f(t + \Delta)\}^{1/2}\, dt.$$

Since by the Stirling formula for large factorials

$$\frac{(2m)!}{m!m!}\left(\frac{1}{2}\right)^{2m} \approx \frac{1}{\sqrt{m\pi}}$$

it follows that

$$\rho(\theta_1, \theta_2) \approx \left\{4H^{1/2}(0)\right\}^m \frac{2m + 1}{\sqrt{m\pi}} \int_{-\infty}^{\infty} e^{-\sigma^2 mt^2/4} \{f(t - \Delta)f(t + \Delta)\}^{1/2}\, dt$$
$$\approx \{4F(\Delta)(1 - F(\Delta))\}^m \cdot 4f(\Delta)/\sigma$$
$$\approx \left\{2\sqrt{F(\Delta)(1 - F(\Delta))}\right\}^n \cdot \frac{2f(\Delta)}{\sigma\{F(\Delta)(1 - F(\Delta))\}^{1/2}}$$
$$= \left\{2\sqrt{F(\Delta)(1 - F(\Delta))}\right\}^n \{Q(\Delta)\}^{-1/2} \qquad \text{(say)}.$$

Then we have

$$Q(\Delta) = \frac{F^2(\Delta) + (1 - F(\Delta))^2}{2F(\Delta)(1 - F(\Delta))} - \frac{f'(\Delta)}{2f^2(\Delta)}(1 - 2F(\Delta)),$$

hence the amount of information between two distributions for $T = \text{med} X_i$ is given by

$$I_T(1,2) = -4n \log(4F(\Delta)(1 - F(\Delta))) + 4 \log Q(\Delta) + o(1).$$

On the other hand the amount of information between two distributions for each X_i is given by

$$I(1,2) = -8 \log \int_{-\infty}^{\infty} f^{1/2}(x - \Delta) f^{1/2}(x + \Delta) dx$$

$$= I(\Delta) \qquad \text{(say)}.$$

Then the asymptotic relative efficiency of $T = \text{med} X_i$ w.r.t. $(X_1, \ldots, X_n)$ is obtained by

(3.4.10) $$\frac{I_T(1,2)}{nI(1,2)} = -\frac{4}{I(\Delta)} \log(4F(\Delta)(1 - F(\Delta))) + o(1).$$

We also have for small $\Delta > 0$

(3.4.11) $$I(\Delta) = -8 \log \int_{-\infty}^{\infty} f^{1/2}(x - \Delta) f^{1/2}(x + \Delta) dx$$

$$\approx -8 \log \int_{-\infty}^{\infty} f(x) \left\{ 1 - \left(\frac{f'(x)}{f(x)} \right)^2 \Delta^2 + \frac{f''(x)}{f(x)} \Delta^2 \right\}^{1/2} dx$$

$$\approx 4\Delta^2 \int_{-\infty}^{\infty} \frac{\{f'(x)\}^2}{f(x)} dx$$

$$= 4\Delta^2 I,$$

provided that $\lim_{|x| \to \infty} f'(x) = 0$, where I denotes the Fisher information of f. Further we obtain for small $\Delta > 0$

(3.4.12) $$4F(\Delta)(1 - F(\Delta)) = 1 - 4 \left(F(\Delta) - \frac{1}{2} \right)^2$$

$$\approx 1 - 4f^2(0)\Delta^2.$$

Hence it follows from (3.4.10) to (3.4.12) that the asymptotic relative efficiency of $T = \text{med} X_i$ w.r.t. $(X_1, \ldots, X_n)$ is given by $4f^2(0)/I$, which is equal to the Pitman efficiency. In particular, letting $f(x) = e^{-|x|}/2$ we have for $\Delta > 0$

$$F(\Delta) = 1 - \frac{1}{2} e^{-\Delta}$$

and

$$I(\Delta) = -8\log \int_{-\infty}^{\infty} \frac{1}{2} e^{-\{|x-\Delta|+|x+\Delta|\}/2} dx$$

$$= -8\log \left(\int_0^{\Delta} e^{-\Delta} dx + \int_{\Delta}^{\infty} e^{-x} dx \right)$$

$$= -8\log \left(e^{-\Delta}(1+\Delta) \right)$$

$$= 8\left(\Delta - \log(1+\Delta) \right).$$

Hence we see that the asymptotic relative efficiency of $T = \mathrm{med} X_i$ w.r.t. $(X_1, \ldots, X_n)$ is given by

$$\frac{\Delta - \log\left(2 - e^{-\Delta}\right)}{2\left\{\Delta - \log(1+\Delta)\right\}},$$

which tends to 1 as $\Delta \to 0$.

3.5. ORDER OF CONSISTENCY

In this section, following Akahira (1994a), we obtain a necessary condition on the amount (3.3.1) of information for the existence of a $\{c_n\}$-consistent estimator, and get a bound for the order of convergence of consistent estimators in such a location parameter family of non-regular densities as in Akahira (1975a) and Akahira and Takeuchi (1981). Then the amount of information plays an important role in the discussion on the order of consistency.

Let $\mathcal{X}$ be an abstract space the element of which is denoted by x. Let $\mathcal{B}$ a σ-field of subsets of $\mathcal{X}$. Let Θ be a parameter space, which is assumed to be an open subset of Euclidean p-space $\mathbf{R}^p$ with a norm denoted by $\|\cdot\|$. We denote by $(\mathcal{X}^{(n)}, \mathcal{B}^{(n)})$ the n-fold direct products of $(\mathcal{X}, \mathcal{B})$. For each $n = 1, 2, \ldots$, the points of $\mathcal{X}^{(n)}$ is denoted by $\mathbf{x} = (x_1, \ldots, x_n)$. We consider a sequence of classes of probability measures $\{P_{\theta,n} : \theta \in \Theta\}$ $(n = 1, 2, \ldots)$ each defined on $(\mathcal{X}^{(n)}, \mathcal{B}^{(n)})$ and where for each n and each $\theta \in \Theta$ the following holds :

$$P_{\theta,n}(B^{(n)}) = P_{\theta,n+1}(B^{(n)} \times \mathcal{X})$$

for all $B^{(n)} \in \mathcal{B}^{(n)}$. An estimator of θ is defined to be a sequence $\{\hat{\theta}_n\}$ where $\hat{\theta}_n$ is a $\mathcal{B}^{(n)}$-measurable function from $\mathcal{X}^{(n)}$ into Θ $(n = 1, 2, \ldots)$. For simplicity we denote an estimator as $\hat{\theta}_n$ instead of $\{\hat{\theta}_n\}$.

Definition 3.5.1. For an increasing sequence of positive numbers $\{c_n\}$ (c_n tending to infinity) an estimator $\hat{\theta}_n$ of θ is called consistent with order $\{c_n\}$ (or $\{c_n\}$-consistent for short) if for every η of Θ, there exist a sufficiently small positive number δ satisfying the following :

$$\lim_{L \to \infty} \varlimsup_{n \to \infty} \sup_{\theta \,:\, \|\theta - \eta\| < \delta} P_{\theta,n} \left\{ c_n \|\hat{\theta}_n - \theta\| \geq L \right\} = 0$$

(Akahira (1975a), Akahira and Takeuchi (1981)).

It is noted that, in the above definition, the condition of the local uniformity on θ is necessary to avoid a phenomenon on the order like "superefficiency." Order $\{c_n\}$ is called to be greater than order $\{c'_n\}$ if $\lim_{n \to \infty} c'_n/c_n = 0$. For any two points θ and θ' in

Θ there exists a σ-finite measure μ_n such that $P_{\theta,n}$ and $P_{\theta',n}$ are absolutely continuous with respect to μ_n. Then for any points θ and θ' in Θ we define
(3.5.1)
$$d_n(\theta,\theta') = \int_{\mathcal{X}^{(n)}} \left| \frac{dP_{\theta,n}}{d\mu_n} - \frac{dP_{\theta',n}}{d\mu_n} \right| d\mu_n = 2 \sup_{B^{(n)} \in \mathcal{B}^{(n)}} \left| P_{\theta,n}\left(B^{(n)}\right) - P_{\theta',n}\left(B^{(n)}\right) \right|.$$

It is easily seen that, for each n, d_n is a metric on Θ which is independent of μ_n.

We also consider the quantity which was defined in Section 3.3 as an amount of information between $P_{\theta,n}$ and $P_{\theta',n}$ for any disjoint points θ and θ' in Θ as follows.

$$(3.5.2) \qquad\qquad I_n(\theta,\theta') = -8 \log \int_{\mathcal{X}^{(n)}} \left(\frac{dP_{\theta,n}}{d\mu_n} \cdot \frac{dP_{\theta',n}}{d\mu_n} \right)^{1/2} d\mu_n.$$

A necessary condition for the existence of a $\{c_n\}$-consistent estimator was given as follows.

Theorem 3.5.1. *If there exists a $\{c_n\}$-consistent estimator of θ, then for each $\theta \in \Theta$ and every $\varepsilon > 0$ there is a positive number t_0 such that, for any $t \geq t_0$,*

$$\varliminf_{n\to\infty} d_n(\theta, \theta - tc_n^{-1}\mathbf{1}) \geq 2 - \varepsilon,$$

where $\mathbf{1} = (1,\ldots,1)'$.

The proof is omitted since it is similar to that in Akahira (1975a) and Akahira and Takeuchi (1981). From (3.5.1) and (3.5.2) we have the following relation of the two distances between two distributions $P_{\theta,n}$ and $P_{\theta',n}$, i.e. $d_n(\theta,\theta')$ and

$$(3.5.3) \qquad \rho_n(\theta,\theta') := 1 - \exp\left\{ -\frac{1}{8} I_n(\theta,\theta') \right\}$$

$$= 1 - \int_{\mathcal{X}^{(n)}} \left(\frac{dP_{\theta,n}}{d\mu_n} \cdot \frac{dP_{\theta',n}}{d\mu_n} \right)^{1/2} d\mu_n$$

$$= \frac{1}{2} \int_{\mathcal{X}^{(n)}} \left\{ \left(\frac{dP_{\theta,n}}{d\mu_n} \right)^{1/2} - \left(\frac{dP_{\theta',n}}{d\mu_n} \right)^{1/2} \right\}^2 d\mu_n.$$

Lemma 3.5.1. *For any points θ and θ' in Θ*

$$2\rho_n(\theta,\theta') \leq d_n(\theta,\theta') \leq 2\sqrt{2\rho_n(\theta,\theta')}.$$

Proof. Denote $g_{\theta,n} := (dP_{\theta,n}/d\mu_n)^{1/2}$ and $g_{\theta',n} := (dP_{\theta',n}/d\mu_n)^{1/2}$. Note that $g_{\theta,n}$ and $g_{\theta',n}$ are non-negative valued, hence $|g_{\theta,n} - g_{\theta',n}| \leq g_{\theta,n} + g_{\theta',n}$, and we have from (3.5.3)

$$d_n(\theta,\theta') = \int_{\mathcal{X}^{(n)}} |g_{\theta,n}^2 - g_{\theta',n}^2| \, d\mu_n = \int_{\mathcal{X}^{(n)}} |g_{\theta,n} - g_{\theta',n}|(g_{\theta,n} + g_{\theta',n}) \, d\mu_n$$

$$\geq \int_{\mathcal{X}^{(n)}} |g_{\theta,n} - g_{\theta',n}|^2 \, d\mu_n = 2\rho_n(\theta,\theta'),$$

and also

$$d_n(\theta, \theta') = \int_{\mathcal{X}^{(n)}} |g_{\theta,n} - g_{\theta',n}| (g_{\theta,n} + g_{\theta',n}) \, d\mu_n$$

$$\leq \left\{ \int_{\mathcal{X}^{(n)}} |g_{\theta,n} - g_{\theta',n}|^2 \, d\mu_n \right\}^{1/2} \left\{ \int_{\mathcal{X}^{(n)}} (g_{\theta,n} + g_{\theta',n})^2 \, d\mu_n \right\}^{1/2}$$

$$\leq \sqrt{2\rho_n(\theta, \theta')} \left\{ 2 \int_{\mathcal{X}^{(n)}} (g_{\theta,n}^2 + g_{\theta',n}^2) \, d\mu_n \right\}^{1/2}$$

$$\leq 2\sqrt{2\rho_n(\theta, \theta')}.$$

Thus we complete the proof.

From Theorem 3.5.1 and Lemma 3.5.1 we have a necessary condition on the amount I_n of information for the existence of a $\{c_n\}$-consistent estimator.

Theorem 3.5.2. *If there exists a $\{c_n\}$-consistent estimator of θ, then for each $\theta \in \Theta$ there exists a positive number t_0 such that, for any $t \geq t_0$,*

$$\lim_{n \to \infty} I_n(\theta, \theta - t c_n^{-1} \mathbf{1}) > 0,$$

where $\mathbf{1} = (1, \ldots, 1)'$.

Proof. From Lemma 3.5.1 we have

$$(3.5.4) \qquad d_n(\theta, \theta - t c_n^{-1} \mathbf{1}) \leq 2\sqrt{2\rho_n(\theta, \theta - t c_n^{-1} \mathbf{1})}$$

$$= 2\sqrt{2} \left[1 - \exp \left\{ -\frac{1}{8} I_n(\theta, \theta - t c_n^{-1} \mathbf{1}) \right\} \right]^{1/2}.$$

Assume that for any $t > 0$ there exists t' such that $t' \geq t$ and $\lim_{n \to \infty} I_n(\theta, \theta - t' c_n^{-1} \mathbf{1}) = 0$. From (3.5.4) we have

$$\overline{\lim_{n \to \infty}} \, d_n(\theta, \theta - t' c_n^{-1} \mathbf{1}) \leq 2\sqrt{2} \lim_{n \to \infty} \left[1 - \exp \left\{ -\frac{1}{8} I_n(\theta, \theta - t' c_n^{-1} \mathbf{1}) \right\} \right]^{1/2} = 0.$$

From Theorem 3.5.1 we see that there does not exist a $\{c_n\}$-consistent estimator of θ. This completes the proof.

Next we consider the bound for the order of consistency. Now let $\mathcal{X} = \Theta = \mathbf{R}^1$, and we suppose that every probability measure $P_\theta(\cdot)$ ($\theta \in \Theta$) is absolutely continuous with respect to the Lebesgue measure and constitutes a location parameter family. Then we denote the density dP_θ/dx by $f(x, \theta)$ and $f(x, \theta) = f_0(x - \theta)$. Let $X_1, \ldots, X_n$ be independent and identically distributed (i.i.d.) random variables with the density $f_0(x - \theta)$. Then we have $P_{\theta,n} = P_\theta^n$, where P_θ^n denotes the n-fold direct product of P_θ. We assume the following condition.

$$(A.3.5.1) \qquad\qquad \begin{aligned} f_0(x) &> 0 \qquad \text{for} \quad a < x < b, \\ f_0(x) &= 0 \qquad \text{for} \quad x \leq a, \, x \geq b. \end{aligned}$$

We consider the amount (3.5.2) of information for $n = 1$, i.e.

(3.5.5) $I(\theta, \theta - \Delta) = -8 \log \int_a^{b-\Delta} \{f_0(x)f_0(x + \Delta)\}^{1/2} dx$

for $0 < \Delta < b - a$. Note that $I_n(\theta, \theta - \Delta) = nI(\theta, \theta - \Delta)$ for $0 < \Delta < b - a$.

For $0 < \Delta < b - a$, putting $g(x) = \{f_0(x) + f_0(x + \Delta)\}/2$, we have

$$g(x) > 0 \quad \text{for} \quad a - \Delta < x < b,$$
$$g(x) = 0 \quad \text{otherwise,}$$

and $\int_{a-\Delta}^b g(x)dx = 1$. Now we put

$$J = \int_{a-\Delta}^b \frac{\{f_0(x + \Delta) - g(x)\}^2}{g(x)} dx.$$

It is noted that the relationship between $d_n(\theta, \theta - \Delta)$ and J is given by

(3.5.6) $d_n(\theta, \theta - \Delta) \leq 2\{(1 + J)^n - 1\}^{1/2},$

which is used to obtain the bound for the order consistency in Akahira (1975a) and Akahira and Takeuchi (1981, Section 2.5). The relationship between J and $I(\theta, \theta - \Delta)$ is given as follows.

Lemma 3.5.2. *Let $f_0(x)$ be a density satisfying (A.3.5.1). For any $\theta \in \Theta$ and $0 < \Delta < b - a$,*

$$0 \leq I(\theta, \theta - \Delta) \leq -8 \log(1 - J).$$

Proof. Since $\int_{a-\Delta}^b g(x)dx = 1$, it follows that

(3.5.7) $0 \leq 1 - \int_a^{b-\Delta} \{f_0(x)f_0(x + \Delta)\}^{1/2} dx$

$$= \int_{a-\Delta}^b g(x)dx - \int_{a-\Delta}^b \{f_0(x)f_0(x + \Delta)\}^{1/2} dx$$

$$= \int_{a-\Delta}^b \left[\frac{1}{2}\{f_0(x) + f_0(x + \Delta)\} - \{f_0(x)f_0(x + \Delta)\}^{1/2}\right] dx$$

$$= \int_{a-\Delta}^b \frac{1}{2} \left\{f_0^{1/2}(x) - f_0^{1/2}(x + \Delta)\right\}^2 dx$$

$$= \frac{1}{2} \int_{a-\Delta}^b \left\{\frac{f_0(x) - f_0(x + \Delta)}{f_0^{1/2}(x) + f_0^{1/2}(x + \Delta)}\right\}^2 dx$$

$$\leq \frac{1}{2} \int_{a-\Delta}^b \frac{\{f_0(x) - f_0(x + \Delta)\}^2}{f_0(x) + f_0(x + \Delta)} dx$$

$$= \int_{a-\Delta}^b \frac{\{f_0(x + \Delta) - g(x)\}^2}{g(x)} dx = J.$$

Since

$$J = \int_{a-\Delta}^{b} \frac{\{f_0(x+\Delta) - g(x)\}^2}{g(x)} dx = \int_{a-\Delta}^{b} \frac{f_0^2(x+\Delta)}{g(x)} dx - 1$$

$$\leq \int_{a-\Delta}^{b-\Delta} 2f_0(x+\Delta) dx - 1 \leq 1,$$

it follows from (3.5.7) that $0 \leq J \leq 1$. From (3.5.7) we also have

$$0 \leq 1 - J \leq \int_a^{b-\Delta} \{f_0(x)f_0(x+\Delta)\}^{1/2} dx \leq 1,$$

hence

$$0 \leq -8 \log \int_a^{b-\Delta} \{f_0(x)f_0(x+\Delta)\}^{1/2} dx = I(\theta, \theta - \Delta) \leq -8 \log(1 - J).$$

This completes the proof.

Further we assume the following conditions.
(A.3.5.2) $f_0(x)$ is twice continuously differentiable in the open interval (a, b) and

$$\lim_{x \to a+0} (x-a)^{1-\alpha} f_0(x) = A',$$

$$\lim_{x \to b-0} (b-x)^{1-\beta} f_0(x) = B',$$

where both α and β are positive constants satisfying $\alpha \leq \beta < \infty$ and A' and B' are positive finite numbers.
(A.3.5.3) $A'' = \lim_{x \to a+0} (x-a)^{2-\alpha} |f_0'(x)|$ and $B'' = \lim_{x \to b-0} (b-x)^{2-\beta} |f_0'(x)|$ are finite. For $\alpha \geq 2$, $f_0''(x)$ are bounded.
For example we see that the Beta distributions $Be(\alpha, \beta)$ $(0 < \alpha \leq \beta \leq 2$ or $3 \leq \alpha \leq \beta < \infty)$ satisfy the conditions (A.3.5.1) to (A.3.5.3).

Then the orders of the value of J are given as follows.

Lemma 3.5.3. *Assume that the conditions (A.3.5.1) to (A.3.5.3) hold. For a small positive number Δ*

$$J = \begin{cases} O(\Delta^\alpha) & \text{for} \quad 0 < \alpha < 2, \\ O(\Delta^2 |\log \Delta|) & \text{for} \quad \alpha = 2, \\ O(\Delta^2) & \text{for} \quad \alpha > 2. \end{cases}$$

The proof is omitted since it is given in Akahira (1975a) and Akahira and Takeuchi (1981). From Lemmas 3.5.2 and 3.5.3 we have the following.

Lemma 3.5.4. *Assume that the conditions (A.3.5.1) to (A.3.5.3) hold. For any $\theta \in \Theta$ and a small positive number $\triangle$*

$$I(\theta, \theta - \triangle) = \begin{cases} O(\triangle^\alpha) & \text{for} \quad 0 < \alpha < 2, \\ O(\triangle^2 |\log \triangle|) & \text{for} \quad \alpha = 2, \\ O(\triangle^2) & \text{for} \quad \alpha > 2. \end{cases}$$

Proof. Since, by Lemma 3.5.3,

$$1 - J = \begin{cases} 1 - O(\triangle^\alpha) & \text{for} \quad 0 < \alpha < 2, \\ 1 - O(\triangle^2 |\log \triangle|) & \text{for} \quad \alpha = 2, \\ 1 - O(\triangle^2) & \text{for} \quad \alpha > 2, \end{cases}$$

it follows that

$$-8\log(1 - J) = \begin{cases} O(\triangle^\alpha) & \text{for} \quad 0 < \alpha < 2, \\ O(\triangle^2 |\log \triangle|) & \text{for} \quad \alpha = 2, \\ O(\triangle^2) & \text{for} \quad \alpha > 2. \end{cases}$$

From Lemma 3.5.2 we obtain

$$I(\theta, \theta - \triangle) = \begin{cases} O(\triangle^\alpha) & \text{for} \quad 0 < \alpha < 2, \\ O(\triangle^2 |\log \triangle|) & \text{for} \quad \alpha = 2, \\ O(\triangle^2) & \text{for} \quad \alpha > 2. \end{cases}$$

This completes the proof.

From Theorem 3.5.2 and Lemma 3.5.4 we have the bound for the order of consistency as follows.

Theorem 3.5.3. *Suppose that $X_1, X_2, \ldots, X_n, \ldots$ is a sequence of i.i.d. random variables with a density function $f_0(x)$ satisfying (A.3.5.1) to (A.3.5.3). Then there does not exist a consistent estimator with order greater than value c_n^* as given as follows.*

$$c_n^* = \begin{cases} n^{1/\alpha} & \text{for} \quad 0 < \alpha < 2, \\ \sqrt{n \log n} & \text{for} \quad \alpha = 2, \\ \sqrt{n} & \text{for} \quad \alpha > 2. \end{cases}$$

Proof. From Lemma 3.5.4 we obtain for sufficiently large n and every $t > 0$

$$nI(\theta, \theta - tc_n^{-1}) = \begin{cases} O(t^\alpha n c_n^{-\alpha}) & \text{for} \quad 0 < \alpha < 2, \\ O(t^2 n c_n^{-2} |\log tc_n^{-1}|) & \text{for} \quad \alpha = 2, \\ O(t^2 n c_n^{-2}) & \text{for} \quad \alpha > 2. \end{cases}$$

(i) $0 < \alpha < 2$. If order c_n is greater than order $n^{1/\alpha}$, then $\lim_{n\to\infty} nI(\theta, \theta - tc_n^{-1}) = 0$ for all $t > 0$ and all $\theta \in \Theta$. Hence it follows from Theorem 3.5.2 that there does not exist a consistent estimator with the order greater than order $n^{1/\alpha}$.

(ii) $\alpha = 2$. If order c_n is greater than order $\sqrt{n \log n}$, then $\lim_{n \to \infty} n I(\theta, \theta - t c_n^{-1}) = 0$ for all $t > 0$ and all $\theta \in \Theta$. Hence it follows from Theorem 3.5.2 that there does not exist a consistent estimator with the order greater than order $\sqrt{n \log n}$.

(iii) $\alpha > 2$. If order c_n is greater than $\sqrt{n}$, then $\lim_{n \to \infty} n I(\theta, \theta - t c_n^{-1}) = 0$ for all $t > 0$ and all $\theta \in \Theta$. Hence it follows from Theorem 3.5.2 that there does not exist a consistent estimator with the order greater than order $\sqrt{n}$.

Thus we complete the proof.

Remark 3.5.1. On the attainment of the bound c_n^* for the order of consistency in Theorem 3.5.3, we can obtain $\{c_n^*\}$-consistent estimator as follows.

α	order c_n^*	$\{c_n^*\}$-consistent estimator
$0 < \alpha < 2$	$n^{1/\alpha}$	$\left\{ \min\limits_{1 \le i \le n} X_i + \max\limits_{1 \le i \le n} X_i - (a+b) \right\} \Big/ 2$
$\alpha = 2$	$\sqrt{n \log n}$	$\hat{\theta}_{ML}$
$\alpha > 2$	$\sqrt{n}$	$\hat{\theta}_{ML}$

Here $\hat{\theta}_{ML}$ is the maximum likelihood estimator of θ (see Akahira (1975a) and Akahira and Takeuchi (1981)).

It is noted that the result of Theorem 3.5.3 coincides with that derived from (3.5.6) and Lemma 3.5.3 which is obtained in Akahira (1975a) and Akahira and Takeuchi (1981). Antoch (1984) also gives some Monte Carlo results on the behaviour of the above estimators and shows that they are fully consistent with the theoretical conclusion. Further the related results to this section are found in Akahira (1975b), Akahira and Takeuchi (1991b) Boente and Fraiman (1988), Hall (1982), Janssen and Reiss (1988), Jurečková (1981), Smith (1985, 1989), Vostrikova (1984), Weiss and Wolfowitz (1973, 1974), Woodroofe (1972, 1974) and others, and the asymptotic sufficiency is discussed by Akahira (1976b), Mita (1979), Weiss (1979) and others in a similar situation.

CHAPTER 4

LOSS OF INFORMATION ASSOCIATED WITH THE ORDER STATISTICS AND RELATED ESTIMATORS IN THE CASE OF DOUBLE EXPONENTIAL DISTRIBUTIONS

4.1. LOSS OF INFORMATION OF THE ORDER STATISTICS

Fisher (1934), starting from his fundamental paper (1922), discussed estimators of the location parameter of a double exponential (two-sided exponential) distribution as a typical example of non-regular estimation. He showed that the maximum likelihood estimator (MLE), which is equal to the sample median in this case, has asymptotic loss of information of order $\sqrt{n}$, as compared to constant order in regular cases. Let I and I_T be the amounts of Fisher information in a single observation and that in a statistic T, respectively. Then the value of $nI - I_T$ as $n \to \infty$, i.e. $\lim_{n \to \infty}(nI - I_T)$ is called the loss of information associated with T and its asymptotic value as $n \to \infty$ is called the asymptotic loss of information (see, e.g. Rao (1961)).

Suppose that $X_1, \ldots, X_n$ are independent and identically distributed (i.i.d.) random variables with the double exponential density function

$$f_0(x - \theta) = \frac{1}{2}\exp(-|x - \theta|).$$

First we compute the amount of information contained in the order statistics around the median, or equivalently the loss of information by discarding all other order statistics. We use the following lemma which applies to any distribution.

Lemma 4.1.1. (Fisher, 1925; Rao, 1961) *The loss of information associated with a statistic T, which is a function of a sample of size n obtained from the population with a density function $f(x, \theta)$, is given by*

$$E_\theta\left[V_\theta\left(\sum_{i=1}^n \frac{\partial}{\partial\theta}\log f(X_i, \theta)\;\middle|\;T\right)\right],$$

provided that the differentiation under the integral signs of the expectations on the distributions of $(X_1, \ldots, X_n)$ and T is allowed, where $E_\theta(\cdot)$ denotes the unconditional expectation and $V_\theta(\cdot \mid T)$ the conditional variance given T.

Outline of the proof.

$$E_\theta\left[V_\theta\left(\sum_{i=1}^n \frac{\partial \log f(X_i,\theta)}{\partial \theta}\middle| T\right)\right]$$

$$=E_\theta\left[E_\theta\left[\left\{\sum_{i=1}^n \frac{\partial \log f(X_i,\theta)}{\partial \theta}\right\}^2\middle| T\right] - \left\{E_\theta\left[\sum_{i=1}^n \frac{\partial \log f(X_i,\theta)}{\partial \theta}\middle| T\right]\right\}^2\right]$$

$$=V_\theta\left(\sum_{i=1}^n \frac{\partial \log f(X_i,\theta)}{\partial \theta}\right) - V_\theta\left(E_\theta\left[\frac{\partial \log f(X_i,\theta)}{\partial \theta}\middle| T\right]\right)$$

$$=nI(\theta) - I_T(\theta)$$

$$=nI(\theta) - V_\theta\left(\frac{\partial \log g(T,\theta)}{\partial \theta}\right).$$

Here $g(\cdot,\theta)$ is a density of the statistic T.

Now let T_k be the set of the central $2k+1$ order statistics $X_{(s-k+1)} \leq \cdots \leq X_{(s+k+1)}$ obtained from a sample of size $n = 2s+1$ from the double exponential distribution with a density $f_0(x - \theta)$. Then we have the following lemma.

Lemma 4.1.2. *For each $k = 0, 1, \ldots, s-1$ the loss of information L_k associated with T_k is given by*

$$(4.1.1) \qquad L_k = E_\theta\left[V_\theta\left(\sum_{i=1}^{2s+1} \mathrm{sgn}(X_i - \theta)\middle| T_k\right)\right]$$

$$= \frac{2(2s+1)!}{(s-k-1)!(s+k)!}\int_{1/2}^1 (2u^{s-k-1} - u^{s-k-2})(1-u)^{s+k}du.$$

Proof. Equation (4.1.1) follows directly from Lemma 4.1.1 and the assumed form of the density function. Given T_k, the set of the first $s-k$ order statistics $X_{(1)}, \ldots, X_{(s-k)}$ is a random ordered sample from the distribution with a density $f_0(x)/F_0(x_{(s-k+1)})$ on $x \leq x_{(s-k+1)}$, where $F_0(x)$ is the double exponential distribution function of the X_i. Similarly, the last $s-k$ are from $f_0(x)/[1 - F_0(x_{(s+k+1)})]$ on $x \geq x_{(s+k+1)}$. Define $u = F_0(X_{(s-k+1)})$, $v = 1 - F_0(X_{(s+k+1)})$, $w_i = F_0(X_{(i)})$, $z_i = 1 - F_0(X_{(i)})$. Then, ignoring the ordering of the first and last $(s-k)$ Xs, $w_1, \ldots, w_{s-k}$ and $z_{s+k+2}, \ldots, z_{2s+1}$ are random samples from uniform distributions on $[0, u]$ and $[0, v]$ respectively. Given T_k, each $\mathrm{sgn}(X_i - \theta)$ is determined for $s - k + 1 \leq i \leq s + k + 1$, so we have

$$V_\theta\left(\sum_{i=1}^{2s+1} \mathrm{sgn}(X_i - \theta)\middle| T_k\right) = (s - k)(V_1 + V_2),$$

where $V_1 = 1 - \left\{ \int_0^u \operatorname{sgn}(x - \theta) dw / u \right\}^2$, which corresponds to the conditional variance of X for $X \leq X_{(s-k+1)}$, V_2 corresponds to that of X for $X \geq X_{(s+k+1)}$, and

$$w = F_0(x - \theta) = \begin{cases} \dfrac{1}{2} e^{-|x-\theta|} & \text{for} \quad x \leq \theta, \\ 1 - \dfrac{1}{2} e^{-|x-\theta|} & \text{for} \quad x > \theta. \end{cases}$$

Then we get

$$V_1 = \begin{cases} 0 & \text{for} \quad X_{(s-k+1)} \leq \theta, \\ (2u - 1)/u^2 & \text{for} \quad X_{(s-k+1)} > \theta, \end{cases}$$

$$V_2 = \begin{cases} 0 & \text{for} \quad X_{(s+k+1)} \geq \theta, \\ (2v - 1)/v^2 & \text{for} \quad X_{(s+k+1)} < \theta, \end{cases}$$

and hence

$$L_k = (s - k)\{E(V_1) + E(V_2)\}.$$

Symmetry now implies that

$$E(V_1) = E(V_2) = \frac{(2s + 1)!}{(s - k)!(s + k)!} \int_{1/2}^1 \frac{(2u - 1)}{u^2} u^{s-k}(1 - u)^{s+k} du,$$

and from this result equation (4.1.1) follows as required.

This result can also be expressed in the following form.

Theorem 4.1.1. For each $k = 0, 1, \ldots, s - 2$, the loss of information L_k associated with T_k is given by

$$(4.1.2) \qquad \frac{L_k}{2(2s + 1)} = \binom{2s}{s} \left(\frac{1}{2}\right)^{2s} - \frac{k + 1}{s - k - 1} + \sum_{j=0}^k \frac{2(k - j + 1)}{s - k - 1} \binom{2s}{s - j} \left(\frac{1}{2}\right)^{2s}.$$

Proof. The essence of the proof lies in recognizing the integrals in (4.1.1) as incomplete beta functions, which in turn are tail probabilities of binomial random variables. Specifically, using $Y(n)$ to denote a symmetric binomial random variable with parameter n, i.e., $P\{Y(n) = p\} = \binom{n}{p}(1/2)^n$, we have

$$\int_{1/2}^1 \frac{n!}{(p - 1)!(n - p)!} x^{p-1}(1 - x)^{n-p} dx = P\{Y(n) \leq p - 1\}$$

for integers $p = 1, \ldots, n$. Then (4.1.1) gives

$$\frac{L_k}{2(2s + 1)} = 2P\{Y(2s) \leq s - k - 1\} - \frac{2sP\{Y(2s - 1) \leq s - k - 2\}}{s - k - 1}.$$

Using symmetry twice, this equals

(4.1.3) $-P\{|Y(2s) - s| \le k\} + \dfrac{sP\{s - k - 1 \le Y(2s - 1) \le s + k\} - k - 1}{s - k - 1}$

$= P\{Y(2s) = s\} - \dfrac{k + 1}{s - k - 1}$

$- 2\sum_{j=0}^{k} \left[P\{Y(2s) = s - j\} - \dfrac{sP\{Y(2s - 1) = s - 1 - j\}}{s - k - 1} \right].$

Since

$$sP\{Y(2s - 1) = i - 1\} = iP\{Y(2s) = i\},$$

the sum in (4.1.3) equals

$$2\sum_{j=0}^{k} \left(1 - \frac{s - j}{s - k - 1}\right) P\{Y(2s) = s - j\}.$$

Substitution in (4.1.3) completes the proof of (4.1.2).

Remark 4.1.1. For the case $k = 0$, Theorem 4.1.1 yields

$$\frac{L_0}{2(2s + 1)} = \frac{1}{s - 1}\left[(s + 1)\binom{2s}{s}\left(\frac{1}{2}\right)^{2s} - 1\right].$$

This expression is consistent with a result in Fisher (1934, p.300).

4.2. THE ASYMPTOTIC LOSS OF INFORMATION

In Akahira and Takeuchi (1981, p.97) we showed that the MLE does not have second order asymptotic efficiency, unlike regular cases. This section extends these results by showing that it is possible to construct an estimator which is asymptotically better than the MLE both in terms of asymptotic variance and asymptotic loss of information in the second order, but that it is impossible to have an estimator which is uniformly better than the MLE in the second order expansion of the distribution. We thus conclude that *there is no second order asymptotically efficient estimator.*

Evaluating the three terms in Theorem 4.1.1 for fixed k and large s shows that asymptotically the loss of information is

(4.2.1) $\dfrac{4\sqrt{s}}{\sqrt{\pi}}\left[1 + O\left(\dfrac{1}{s}\right)\right] - 4(k + 1) + O\left(\dfrac{k^2}{\sqrt{s}}\right).$

But in order to improve this quantity we must increase k with s, and (4.2.1) suggests that we take $k = r\sqrt{s} + o(\sqrt{s})$ with a non-negative constant r.

From Theorem 4.1.1 and the Cramér-Rao inequality it follows that for any unbiased estimator $\hat{\theta}$ which is a function of T_k only, we have for fixed k

(4.2.2) $$V_\theta(\hat{\theta}) \geq \frac{1}{I_{\hat{\theta}}} \geq \frac{1}{I_{T_k}} = \frac{1}{n - L_k} = \frac{1}{n}\left\{1 + \frac{L_k}{n} + o\left(\frac{L_k}{n}\right)\right\},$$

where $I_{\hat{\theta}}$ and I_{T_k} denote the Fisher information at $\hat{\theta}$ and T_k respectively.

Theorem 4.2.1. For $k = r\sqrt{s} + o(\sqrt{s}) = (1/2)\rho\sqrt{n} + o(\sqrt{n})$ for $\rho = r\sqrt{2} \geq 0$ and $n = 2s + 1$, the loss of information, L_k, associated with T_k is given by

(4.2.3) $$L_k = 4[\phi(\rho) - \rho\{1 - \Phi(\rho)\}]\sqrt{n} + o(\sqrt{n}),$$

where $\Phi(x) = \int_{-\infty}^{x} \phi(u)du$ with $\phi(u) = e^{-u^2/2}/\sqrt{2\pi}$.

Proof. Using the normal approximation to the binomial distribution we have

(4.2.4) $$P\{Y(2s) = s\} = \frac{1}{\sqrt{\pi s}}\left(1 - \frac{1}{8s} + O\left(\frac{1}{s^2}\right)\right).$$

The summation term in (4.1.2) equals

(4.2.5) $$2\sum_{j=0}^{k} \frac{k - j + 1}{s - k - 1}P\{Y(2s) = s + j\}$$

$$= \frac{2(k+1)}{s - k - 1}\sum_{j=0}^{k} P\{Y(2s) = s + j\} - \frac{2}{s - k - 1}\sum_{j=0}^{k} jP\{Y(2s) = s + j\}$$

$$= \frac{2(r\sqrt{s} + 1)}{s - r\sqrt{s} - 1}\int_0^\rho \phi(u)du - \frac{\sqrt{2s}}{s - r\sqrt{s} - 1}\int_0^\rho u\phi(u)du + o\left(\frac{1}{\sqrt{s}}\right)$$

after approximation by Riemann integrals,

$$= \frac{2(r\sqrt{s} + 1)}{s - r\sqrt{s} - 1}\int_0^\rho \phi(u)du - \frac{1 - e^{-r^2}}{\sqrt{2\pi}} \cdot \frac{\sqrt{2\pi}}{s - r\sqrt{s} - 1} + o\left(\frac{1}{\sqrt{s}}\right).$$

By (4.1.2), (4.2.4) and (4.2.5) we obtain

$$L_k = 2(2s + 1)\left\{-\frac{r}{\sqrt{s}} + \frac{2r}{\sqrt{s}}\int_0^\rho \phi(u)du + \frac{1}{\sqrt{\pi s}}e^{-r^2}\right\} + o\left(\sqrt{s}\right)$$

$$= 4\left[\phi(\rho) - \rho\{1 - \Phi(\rho)\}\right]\sqrt{n} + o(\sqrt{n}),$$

which completes the proof.

We now consider an explicit form of estimators depending on T_k. As the simplest one, we put $\hat{\theta}_k = \left(X_{(s-k+1)} + X_{(s+k+1)}\right)/2$. The proof of the next result is in Section 4.3.

Theorem 4.2.2. For $k = r\sqrt{s} + o(\sqrt{s}) = (1/2)\rho\sqrt{n} + o(\sqrt{n})$ where $\rho = r\sqrt{2} \geq 0$ and $n = 2s + 1$, the estimator $\hat{\theta}_k$ has the stochastic expansion

$$\sqrt{n}\left(\hat{\theta}_k - \theta\right) = Z + \frac{(Z-\rho)|Z-\rho| + (Z+\rho)|Z+\rho|}{4\sqrt{n}} + o_p\left(n^{-1/2}\right)$$

and asymptotic variance

(4.2.6) $$V_\theta\left(\sqrt{n}\hat{\theta}_k\right) = 1 + c_\rho n^{-1/2}\left(1 + o(1)\right),$$

where $Z = \sqrt{n}\left\{F_0\left(X_{(s-k+1)} - \theta\right) + F_0\left(X_{(s+k+1)} - \theta\right) - 1\right\}$ with $F_0(\,\cdot\,)$ the distribution function of X_i, equivalently, $Z = \sqrt{n}\left(U_{(s-k+1)} + U_{(s+k+1)} - 1\right)$ with $U_{(1)}, \ldots, U_{(n)}$ the order statistics from the uniform distribution on the interval $(0,1)$, and

(4.2.7) $$c_\rho = 2\left[2\phi(\rho) + \rho\left\{2\Phi(\rho) - \frac{3}{2}\right\}\right].$$

Remark 4.2.1. Since $\phi(x) \geq x\{1 - \Phi(x)\}$ for $x \geq 0$, it follows that, for any $\rho \geq 0$, $c_\rho \geq \rho$. Define ρ_0 by $\Phi(\rho_0) = 3/4$, so that $\rho_0 \approx 0.67$. Then c_ρ has a minimum value at $\rho = \rho_0$. Hence $\hat{\theta}_k$ has minimum variance at $\rho = \rho_0$.

Remark 4.2.2. In particular, for $k = 0$ i.e., $\rho = 0$, we obtain from Theorem 4.2.2,

(4.2.8) $$V_\theta\left(\sqrt{n}\hat{\theta}_0\right) = 1 + \frac{2\sqrt{2}}{\sqrt{\pi n}} + o\left(\frac{1}{\sqrt{n}}\right).$$

Note that $\hat{\theta}_0$ is the median, i.e., the MLE in this case.

Remark 4.2.3. Since

$$V_\theta\left(\hat{\theta}_k\right) = n^{-1}\left[1 + c_\rho n^{-1/2}\{1 + o(1)\}\right],$$

it follows from (4.2.2) that $c_\rho \geq L_k/\sqrt{n}$. In this case we have from (4.2.3) and Theorem 4.2.2,

$$c_\rho - \frac{1}{\sqrt{n}}L_k = \rho.$$

Corollary 4.2.1. For $k = r\sqrt{s} + o(\sqrt{s}) = (1/2)\rho\sqrt{n} + o(\sqrt{n})$ where $\rho = r\sqrt{2} \geq 0$ and $n = 2s + 1$, the bound of the loss of information L'_k associated with the estimator $\hat{\theta}_k$ is given by $L'_k \leq c_{\rho_0}\sqrt{n} + o(\sqrt{n})$, where ρ_0 is defined in Remark 4.2.1.

Proof. From (4.2.2) we have $V_\theta\left(\hat{\theta}_k\right) \geq (n - L'_k)^{-1}$, which implies that

$$L'_k \leq n - \frac{1}{V_\theta\left(\hat{\theta}_k\right)} = n - n\left(1 + \frac{c_\rho}{\sqrt{n}}\right)^{-1} + o(\sqrt{n}) = c_\rho\sqrt{n} + o(\sqrt{n}).$$

Therefore, by putting $\rho = \rho_0$ we have

$$L'_k \leq c_{\rho_0}\sqrt{n} + o(\sqrt{n}).$$

Remark 4.2.4. Since $\rho_0 \approx 0.67$, it is seen that

$$(4.2.9) \qquad L'_k \leq c_{\rho_0}\sqrt{n} + o(\sqrt{n}) = 1.27\sqrt{n} + o(\sqrt{n}).$$

On the other hand, it follows from (4.2.1) that the loss L'_0 of the MLE $\hat{\theta}_0$, i.e., the median, is given by

$$(4.2.10) \qquad L'_0 = L_0 = \frac{2\sqrt{2}}{\sqrt{\pi}}\sqrt{n} + o(\sqrt{n}) \fallingdotseq 1.60\sqrt{n} + o(\sqrt{n}).$$

From (4.2.9) and (4.2.10) we see that the estimator $\hat{\theta}_k$ is asymptotically better than the MLE $\hat{\theta}_0$ in the sense of $L'_k \leq L'_0 + o(\sqrt{n})$.

In Akahira and Takeuchi (1981, p.97) it is shown that the asymptotic distribution of $\hat{\theta}_0$ up to the order $n^{-1/2}$ is given by

$$(4.2.11) \qquad P_\theta^n\left\{\sqrt{n}\left(\hat{\theta}_0 - \theta\right) \leq t\right\} = \Phi(t) - \frac{1}{2}n^{-1/2}t^2\phi(t)\mathrm{sgnt} + o\left(n^{-1/2}\right)$$
$$= F_n(t) \qquad \text{(say)}.$$

Then the asymptotic density $f_n(t)$ of $\hat{\theta}_0$ is given by

$$f_n(t) = F'_n(t) = \phi(t) - \frac{1}{2}n^{-1/2}(2t - t^3)\phi(t)\mathrm{sgnt} + o\left(n^{-1/2}\right).$$

Hence the asymptotic variance of $\hat{\theta}_0$ is given by

$$V_\theta\left(\sqrt{n}\hat{\theta}_0\right) = \int_{-\infty}^{\infty} t^2 f_n(t)dt = 1 + \frac{2\sqrt{2}}{\sqrt{\pi n}}(1 + o(1)),$$

which is consistent with (4.2.8), and identically so when $n = 2s + 1$.

Next we obtain the asymptotic distribution of $\hat{\theta}_k$ up to the order $n^{-1/2}$. The proof is given in Section 4.3.

Theorem 4.2.3. For $k = r\sqrt{s} + o(\sqrt{s}) = (1/2)\rho\sqrt{n} + o(\sqrt{n})$ where $\rho = r\sqrt{2} \geq 0$ and $n = 2s + 1$, the asymptotic distribution of $\hat{\theta}_k$ up to the order $n^{-1/2}$ is given by

$$P_\theta^n\left\{\sqrt{n}\left(\hat{\theta}_k - \theta\right) \leq t\right\}$$
$$= \Phi(t) - \frac{\phi(t)}{2\sqrt{n}}\left[-\rho t + \frac{1}{2}\{(t - \rho)|t - \rho| + (t + \rho)|t + \rho|\}\right] + o\left(\frac{1}{\sqrt{n}}\right).$$

Remark 4.2.5. Letting $r = 0$, we see that, for $k = 0$, the asymptotic distribution of the estimator $\hat{\theta}_0$ is consistent with that given by (4.2.11).

From (4.2.11) and Theorem 4.2.3 we have the following (the proof is straightforward).

Corollary 4.2.2. *For estimators* $\hat{\theta}_k$ *and* $\hat{\theta}_0$,

$$P_\theta^n\left\{\sqrt{n}\left(\hat{\theta}_0 - \theta\right) \le t\right\} \le P_\theta^n\left\{\sqrt{n}\left(\hat{\theta}_k - \theta\right) \le t\right\} + o\left(1/\sqrt{n}\right) \qquad \text{for } t \ge \rho,$$

$$P_\theta^n\left\{\sqrt{n}\left(\hat{\theta}_0 - \theta\right) \le t\right\} \ge P_\theta^n\left\{\sqrt{n}\left(\hat{\theta}_k - \theta\right) \le t\right\} + o\left(1/\sqrt{n}\right) \qquad \text{for } t \le -\rho,$$

$$P_\theta^n\left\{\sqrt{n}\left(\hat{\theta}_0 - \theta\right) \le t\right\} \ge P_\theta^n\left\{\sqrt{n}\left(\hat{\theta}_k - \theta\right) \le t\right\} + o\left(1/\sqrt{n}\right) \qquad \text{for } 0 \le t < \rho,$$

$$P_\theta^n\left\{\sqrt{n}\left(\hat{\theta}_0 - \theta\right) \le t\right\} \le P_\theta^n\left\{\sqrt{n}\left(\hat{\theta}_k - \theta\right) \le t\right\} + o\left(1/\sqrt{n}\right) \qquad \text{for } -\rho < t \le 0.$$

Remark 4.2.6. From Corollary 4.2.2 we see that, for $0 < t < \rho$, $\hat{\theta}_0$ is asymptotically better than $\hat{\theta}_k$ in the sense of the concentration probability. That is,

$$P_\theta^n\left\{\sqrt{n}\left|\hat{\theta}_0 - \theta\right| \le t\right\} \ge P_\theta^n\left\{\sqrt{n}\left|\hat{\theta}_k - \theta\right| \le t\right\} + o(1/\sqrt{n}) \qquad \text{for } 0 < t < \rho,$$

and, for $t \ge \rho$, $\hat{\theta}_k$ is asymptotically better than $\hat{\theta}_0$ in the same sense as the above, i.e.,

$$P_\theta^n\left\{\sqrt{n}\left|\hat{\theta}_0 - \theta\right| \le t\right\} \le P_\theta^n\left\{\sqrt{n}\left|\hat{\theta}_k - \theta\right| \le t\right\} + o(1/\sqrt{n}) \qquad \text{for } t \ge \rho.$$

Recently, Sugiura and Naing (1989) extended the above results to the other estimators using a weighted linear combination of the sample median and pairs of order statistics. Akahira and Takeuchi (1993) also obtained the Bhattacharyya bound for the variance of unbiased estimators and discussed the loss of information of the MLE based on the double exponential distribution rounded off.

4.3. PROOFS OF THEOREMS IN SECTION 4.2

In this section we give the proofs of Theorems 4.2.2 and 4.2.3. The following lemmas are useful for manipulation of the joint distribution of order statistics from the uniform distribution.

Lemma 4.3.1. *Assume that* $U_{(j)}$s *are the order statistics, i.e.* $U_{(1)} \le \cdots \le U_{(n)}$, *of size* n *from the uniform distribution on the interval* $(0,1)$. *Then it is expressed as*

$$U_{(j)} = \sum_{i=1}^{j} Y_i \bigg/ \sum_{i=1}^{n} Y_i \qquad (j = 1, \ldots, n),$$

where $Y_1, \ldots, Y_n$ *are i.i.d. random variables with an exponential density* $f(y, \theta) = e^{-y}$ *for* $y > 0$.

The proof is omitted since it is starightforward.

Lemma 4.3.2. *Let $Y_1, \ldots, Y_n$ be i.i.d. random variables with an exponential density function*

$$(4.3.1) \qquad f(y, \theta) = \begin{cases} e^{-y/\theta}/\theta & \text{for } y > 0, \\ 0 & \text{for } y \leq 0, \end{cases}$$

and let $g(Y_1, \ldots, Y_n)$ be homogeneous in $Y_1, \ldots, Y_n$ of degree 0 so that $g(\kappa Y_1, \ldots, \kappa Y_n) \equiv g(Y_1, \ldots, Y_n)$ for all non-zero κ. Then $g(Y_1, \ldots, Y_n)$ and $\sum_{i=1}^{n} Y_i$ are independent.

Proof. It is clear that $\sum_{i=1}^{n} Y_i$ is a complete sufficient statistic for θ. Since the distribution of $g(Y_1, \ldots, Y_n)$ is independent of θ, then, by the theorem of Basu (1955), it is independent of $\sum_{i=1}^{n} Y_i$.

Lemma 4.3.3. *Under the same conditions as Lemma 4.3.2,*

$$E\left[g(Y_1, \ldots, Y_n)\right] = \frac{E\left[g(Y_1, \ldots, Y_n)\left(\sum_{i=1}^{n} Y_i\right)^{\alpha}\right]}{E\left[\left(\sum_{i=1}^{n} Y_i\right)^{\alpha}\right]}$$

for all real numbers α.

The proof is straightforward from Lemma 4.3.2.

Proof of Theorem 4.2.2. Without loss of generality, we assume that $\theta = 0$. Let $F_0(x)$ be a distribution function of X_i $(i = 1, \ldots, n)$. Then we have $X_{(i)} = F_0^{-1}\left(U_{(i)}\right)$ $(i = 1, \ldots, n)$, where the $U_{(i)}$s are order statistics from the uniform distribution on the interval $(0, 1)$. We also obtain

$$F_0^{-1}(u) = -\left\{\text{sgn}\left(u - \frac{1}{2}\right)\right\} \log(1 - |2u - 1|).$$

We put $U = U_{(s-k+1)}$ and $V = U_{(s+k+1)}$. Then we have

$$n^{1/2}\hat{\theta}_k = \frac{1}{2}n^{1/2}\{F^{-1}(U) + F^{-1}(V)\}$$

$$= n^{1/2}(U + V - 1) + n^{-1/2}\left\{n^{1/2}\left(U - \frac{1}{2}\right)\left|n^{1/2}\left(U - \frac{1}{2}\right)\right|\right\}$$

$$+ n^{-1/2}\left\{n^{1/2}\left(V - \frac{1}{2}\right)\left|n^{1/2}\left(V - \frac{1}{2}\right)\right|\right\} + o_p\left(n^{-1/2}\right).$$

Putting $X = n^{1/2}(U - (1/2))$, $Y = n^{1/2}(V - (1/2))$, we obtain

$$(4.3.2) \qquad n^{1/2}\hat{\theta}_k = X + Y + n^{-1/2}(X|X| + Y|Y|) + o_p\left(n^{-1/2}\right).$$

If $Y_1, \ldots, Y_{2s+1}$ are i.i.d. random variables with the exponential density function (4.3.1) with $\theta = 1$, then it follows from Lemma 4.3.1 that

$$U_{(j)} = \sum_{i=1}^{j} Y_i \bigg/ \sum_{i=1}^{2s+1} Y_i$$

for each $j = 1, \ldots, 2s + 1$. Using this fact we have, by Lemma 4.3.3,

$$
\begin{aligned}
E[(U + V - 1)^2] &= E\left[\left(\frac{\sum_{i=1}^{s-k+1} Y_i - \sum_{i=s+k+2}^{2s+1} Y_i}{\sum_{i=1}^{2s+1} Y_i}\right)^2\right] \\
&= \frac{E\left[\left(\sum_{i=1}^{s-k+1} Y_i - \sum_{i=s+k+1}^{2s+1} Y_i\right)^2\right]}{E\left[\left(\sum_{i=1}^{2s+1} Y_i\right)^2\right]} \\
&= \frac{s - k + 1}{(s + 1)(2s + 1)}.
\end{aligned}
$$

Hence we obtain

$$
V(X + Y) = E[(X + Y)^2] = E\left[\{\sqrt{n}(U + V - 1)\}^2\right]
$$

(4.3.3)
$$
= \frac{s - k + 1}{s + 1} = 1 - \frac{\rho}{\sqrt{n}} + o\left(n^{-1/2}\right).
$$

Since

$$
E(X - Y) = E\left[n^{1/2}(U - V)\right] = -\rho,
$$

it follows similarly that

$$
V(X - Y) = E\left[\{\sqrt{n}(U - V)\}^2\right] - \rho^2 = n^{-1/2}(\rho + o(1)).
$$

We put

$$
Z = X + Y ; \qquad W = s^{1/4}(X - Y + \rho).
$$

Since

$$
X = \frac{1}{2}(X + Y + X - Y) = \frac{1}{2}(Z - \rho + s^{-1/4}W) ;
$$
$$
Y = \frac{1}{2}\{(X + Y) - (X - Y)\} = \frac{1}{2}(Z + \rho - s^{-1/4}W),
$$

we have from (4.3.2)

(4.3.4) $n^{1/2}\hat{\theta}_k = Z + \frac{1}{4}n^{-1/2}\{(Z - \rho)|Z - \rho| + (Z + \rho)|Z + \rho|\} + o_p\left(n^{-1/2}\right).$

Hence

(4.3.5) $V\left(\sqrt{n}\hat{\theta}_k\right) = E(Z^2) + \frac{1}{2}n^{-1/2}\left\{E\left[Z(Z - \rho)|Z - \rho|\right]\right.$

$$
\left. + E\left[Z(Z + \rho)|Z + \rho|\right]\right\} + o\left(n^{-1/2}\right).
$$

For any constant c we have

(4.3.6) $E\left[Z(Z - c)|Z - c|\right] = E\left[|Z - c|^3\right] + cE\left[(Z - c)|Z - c|\right].$

We obtain

(4.3.7) $$\int_{-\infty}^{\infty} |x - c|^3 \phi(x)dx = 2(c^2 + c)\phi(c) + (c^3 + 3c)\{2\Phi(c) - 1\},$$

(4.3.8) $$\int_{-\infty}^{\infty} (x - c)|x - c|\phi(x)dx = -2c\phi(c) + (c^2 + 1)\{1 - 2\Phi(c)\}.$$

Since Z is asymptotically normally distributed with mean 0 and variance $1 - r/\sqrt{s} = 1 - \rho/\sqrt{n}$, we have from (4.3.6) to (4.3.8) that

$$E\left[Z(Z - c)|Z - c|\right] = 4\phi(c) + 2c\{2\Phi(c) - 1\}.$$

Hence we obtain

(4.3.9) $$E\left[Z(Z - \rho)|Z - \rho|\right] = 4\phi(\rho) + 2\rho\{2\Phi(\rho) - 1\}.$$

From (4.3.3), (4.3.5) and (4.3.9) we have

$$V\left(n^{1/2}\hat{\theta}_k\right) = 1 + 2n^{-1/2}\left[2\phi(\rho) + \rho\{2\Phi(\rho) - (3/2)\}\right] + o\left(n^{-1/2}\right).$$

The coefficient of $n^{-1/2}$ here yields c_ρ as at (4.2.7), proving (4.2.6).

Proof of Theorem 4.2.3. Assume without loss of generality that $\theta = 0$. From (4.3.4) we have that $P_\theta^n\left\{\sqrt{n}\hat{\theta}_k \leq t\right\}$ equals

$$P_\theta^n\left\{Z + \frac{1}{4}n^{-1/2}\left[(Z - \rho)|Z - \rho| + (Z + \rho)|Z + \rho|\right] \leq t + o_p\left(n^{-1/2}\right)\right\}$$

$$= P_\theta^n\left\{Z \leq t - \frac{1}{4}n^{-1/2}\left[(t - \rho)|t - \rho| + (t + \rho)|t + \rho|\right] + o_p\left(n^{-1/2}\right)\right\}.$$

We see that Z is asymptotically normally distributed with mean 0 and variance $1 + o(1/\sqrt{s}) = 1 + o(1/\sqrt{n})$. Hence we obtain

$$P_\theta^n\left\{\sqrt{n}\hat{\theta}_k \leq t\right\} = P_\theta^n\left\{Z \leq t - \frac{1}{2}n^{-1/2}\left\{-\rho t + \frac{1}{2}[(t - \rho)|t - \rho|\right.\right.$$

$$\left.\left. + (t + \rho)|t + \rho|]\right\} + o_p\left(n^{-1/2}\right)\right\}$$

$$= \Phi(t) - \frac{1}{2}n^{-1/2}\phi(t)\left[\rho t + \text{sgn}(t)\left\{(|t| - \rho)^+\right\}^2\right] + o\left(n^{-1/2}\right),$$

where $(\ )^+$ denotes a positive part of $(\ )$.

4.4. DISCRETIZED LIKELIHOOD ESTIMATION

In this section we obtain the asymptotic distribution of the discretized likelihood estimator up to the second order for the case of the double exponential distribution, and note that the discretized likelihood estimator has the most concentration probability at any fixed point, which is connected with the result by Akahira and Takeuchi (1979, 1981).

Suppose that $X_1, \ldots, X_n$ are independent and identically distributed real random variables with a density function $f(x, \theta)$. Then we consider the case of a symmetric density with a location parameter, i.e., $f(x, \theta) = f_0(x - \theta)$ and $f_0(x) = f_0(-x)$ for all $x \in \mathbf{R}$. The discretized likelihood estimator was defined by Akahira and Takeuchi (1979, 1981) as follows. For any $u > 0$, the solution $\theta = \hat{\theta}_{DL}^u$ of the discretized likelihood equation

$$
(4.4.1) \qquad \prod_{i=1}^{n} f_0\left(X_i - \theta - \frac{u}{\sqrt{n}}\right) - \prod_{i=1}^{n} f_0\left(X_i - \theta + \frac{u}{\sqrt{n}}\right) = 0
$$

is called the discretized likelihood estimator (DLE) of θ.

It is shown by Akahira and Takeuchi (1979, 1981) and Akahira (1986) that the DLE $\hat{\theta}_{DL}^u$ maximizes the concentration probability $P_\theta^n\left\{\sqrt{n}\left|\hat{\theta}_n - \theta\right| \le u\right\}$ up to the order $n^{-1/2}$ in the class of the all second order asymptotically median unbiased estimators $\hat{\theta}_n$ of θ. Here an estimator $\hat{\theta}_n$ of θ is called second order asymptotically median unbiased (AMU) if

$$
\lim_{n \to \infty} \sqrt{n}\left|P_\theta^n\left\{\hat{\theta}_n \le \theta\right\} - \frac{1}{2}\right| = \lim_{n \to \infty} \sqrt{n}\left|P_\theta^n\left\{\hat{\theta}_n \ge \theta\right\} - \frac{1}{2}\right| = 0
$$

uniformly in some neighborhood of θ.

Now we deal with the case of the double exponential distribution with a density function

$$
f_0(x - \theta) = \frac{1}{2}\exp(-|x - \theta|).
$$

We shall obtain the asymptotic distribution of the DLE up to the order $n^{-1/2}$. In order to obtain the asymptotic distribution of the DLE, putting

$$
T_t = \sum_{i=1}^{n} \phi_{x_i}\left(\frac{t}{\sqrt{n}}\right)
$$

with

$$
\phi_x(\theta) = \left\{\log f_0\left(x - \theta - \frac{u}{\sqrt{n}}\right) - \log f_0\left(x - \theta + \frac{u}{\sqrt{n}}\right)\right\}\bigg/ 2,
$$

we have its mean, variance and cumulant as follows.

Lemma 4.4.1. *For sufficiently large* n,

$$
E_\theta(T_t) = -tu + \frac{|t + u|^3 - |t - u|^3}{12\sqrt{n}} + o\left(\frac{1}{\sqrt{n}}\right),
$$

$$
V_\theta(T_t) = u^2 - \frac{2u^3}{3\sqrt{n}} + o\left(\frac{1}{\sqrt{n}}\right),
$$

$$
E_\theta[\{T_t - E(T_t)\}^3] = o\left(\frac{1}{\sqrt{n}}\right).
$$

Proof. Since

$$\phi_x(\theta) = \frac{1}{2}\left(-\left|x - \theta - \frac{u}{\sqrt{n}}\right| + \left|x - \theta + \frac{u}{\sqrt{n}}\right|\right)$$

(4.4.2)
$$= \begin{cases} \dfrac{u}{\sqrt{n}} & \text{for} \quad \theta \le x - \dfrac{u}{\sqrt{n}}, \\[2mm] x - \theta & \text{for} \quad x - \dfrac{u}{\sqrt{n}} \le \theta \le x + \dfrac{u}{\sqrt{n}}, \\[2mm] -\dfrac{u}{\sqrt{n}} & \text{for} \quad \theta \ge x + \dfrac{u}{\sqrt{n}}, \end{cases}$$

it follows from the definition of the DLE $\hat{\theta}_{DL}^u$ that

$$\sum_{i=1}^{n} \phi_{x_i}\left(\hat{\theta}_{DL}^u\right) = 0.$$

Hence it is seen that the event $\{\hat{\theta}_{DL}^u \le \theta + \nu\}$ is equivalent to the event $\{\sum_{i=1}^{n} \phi_{x_i}(\theta + \nu) \le 0\}$ for a small $|\nu|$. Without loss of generality we assume that $\theta = 0$. If $\nu = t/\sqrt{n}$, then

(4.4.3) $$P_0^n\left\{\hat{\theta}_{DL}^u \le \frac{t}{\sqrt{n}}\right\} = P_0^n\left\{\sum_{i=1}^{n} \phi_{X_i}\left(\frac{t}{\sqrt{n}}\right) \le 0\right\} + o\left(\frac{1}{\sqrt{n}}\right).$$

We obtain the mean, variance and cumulant of T_t as follows.

$$E_0(T_t) = nE_0\left[\phi_X\left(\frac{t}{\sqrt{n}}\right)\right],$$

$$V_0(T_t) = nV_0\left(\phi_X\left(\frac{t}{\sqrt{n}}\right)\right),$$

$$E_0\left[\{T_t - E(T_t)\}^3\right] = nE_0\left[\left\{\phi_X\left(\frac{t}{\sqrt{n}}\right) - E_0\left(\phi_X\left(\frac{t}{\sqrt{n}}\right)\right)\right\}^3\right].$$

We also put $a = (t + u)/\sqrt{n}$ and $b = (t - u)/\sqrt{n}$. From (4.4.2) we have

$$E_0\left[\phi_X\left(\frac{t}{\sqrt{n}}\right)\right] = \frac{u}{\sqrt{n}}\left[P_0\{X \ge a\} - P_0\{X \le b\}\right]$$

$$+ E_0\left[X - \frac{t}{\sqrt{n}} \,\middle|\, b \le X \le a\right] P_0\{b \le X \le a\}$$

$$= \frac{u}{\sqrt{n}}\left\{\int_a^\infty f_0(x)dx - \int_{-\infty}^b f_0(x)dx\right\} + \int_b^a\left(x - \frac{t}{\sqrt{n}}\right)f_0(x)dx$$

$$= -\frac{u}{\sqrt{n}} \left\{ \int_0^a f_0(x)dx + \int_0^b f_0(x)dx \right\} + \int_0^a \left(x - \frac{t}{\sqrt{n}} \right) f_0(x)dx$$

$$- \int_0^b \left(x - \frac{t}{\sqrt{n}} \right) f_0(x)dx,$$

$$E_0\left[\left\{ \phi_X\left(\frac{t}{\sqrt{n}}\right) \right\}^2 \right] = \frac{u^2}{n} \left\{ 1 - \int_0^a f_0(x)dx + \int_0^b f_0(x)dx \right\}$$

$$+ \int_0^a \left(x - \frac{t}{\sqrt{n}} \right)^2 f_0(x)dx - \int_0^b \left(x - \frac{t}{\sqrt{n}} \right)^2 f_0(x)dx,$$

$$E_0\left[\left\{ \phi_X\left(\frac{t}{\sqrt{n}}\right) \right\}^3 \right] = \frac{u^3}{n\sqrt{n}} \left\{ \int_0^a f_0(x)dx + \int_0^b f_0(x)dx \right\}$$

$$+ \int_0^a \left(x - \frac{t}{\sqrt{n}} \right)^3 f_0(x)dx - \int_0^b \left(x - \frac{t}{\sqrt{n}} \right)^3 f_0(x)dx.$$

For each $k = 1, 2, \ldots$ and any constant c we obtain

$$\int_0^c x^k f_0(x)dx = \frac{1}{2} \int_0^c x^k e^{-|x|}dx = \frac{1}{2}\left\{ \frac{c^{k+1}}{k+1} - \frac{1}{k+2}(\mathrm{sgn}c)c^{k+2} + \frac{c^{k+3}}{2(k+3)} - \cdots \right\}.$$

Then we have

$$E_0\left[\phi_X\left(\frac{t}{\sqrt{n}}\right) \right] = \frac{u}{\sqrt{n}} \left\{ -\frac{1}{2}\left(a - \frac{\mathrm{sgn}a}{2}a^2 \right) - \frac{1}{2}\left(b - \frac{\mathrm{sgn}b}{2}b^2 \right) \right\}$$

$$+ \frac{1}{2}\left(\frac{a^2}{2} - \frac{\mathrm{sgn}a}{3}a^3 - \frac{b^2}{2} + \frac{\mathrm{sgn}b}{3}b^3 \right)$$

$$- \frac{t}{\sqrt{n}} \left\{ \frac{1}{2}\left(a - \frac{\mathrm{sgn}a}{2}a^2 \right) - \frac{1}{2}\left(b - \frac{\mathrm{sgn}b}{2}b^2 \right) \right\} + o\left(\frac{1}{n\sqrt{n}} \right)$$

$$= -\frac{tu}{n} + \frac{|t+u|^3 - |t-u|^3}{12n\sqrt{n}} + o\left(\frac{1}{n\sqrt{n}} \right).$$

We also obtain

$$E_0\left[\left\{ \phi_X\left(\frac{t}{\sqrt{n}}\right) \right\}^2 \right] = \frac{u^2}{n} - \frac{2u^3}{3n\sqrt{n}} + o\left(\frac{1}{n\sqrt{n}} \right),$$

$$E_0\left[\left\{ \phi_X\left(\frac{t}{\sqrt{n}}\right) \right\}^3 \right] = o\left(\frac{1}{n\sqrt{n}} \right).$$

Hence we have

$$E_0(T_t) = -tu + \frac{|t+u|^3 - |t-u|^3}{12\sqrt{n}} + o\left(\frac{1}{\sqrt{n}}\right),$$

$$V_0(T_t) = u^2 - \frac{2u^3}{3\sqrt{n}} + o\left(\frac{1}{\sqrt{n}}\right),$$

$$E_0\left[\{T_t - E(T_t)\}^3\right] = o\left(\frac{1}{\sqrt{n}}\right).$$

Thus we complete the proof.

From Lemma 4.4.1 we have the following theorem.

Theorem 4.4.1. *The asymptotic distribution of the discretized likelihood estimator, up to the second order, i.e., the order* $n^{-1/2}$, *is given by*

$$P_\theta^n\left\{\sqrt{n}\left(\hat{\theta}_{DL}^u - \theta\right) \leq t\right\}$$

$$= \begin{cases} \Phi(t) - \dfrac{1}{6\sqrt{n}}\phi(t)\dfrac{t(t^2+u^2)}{u} + o\left(\dfrac{1}{\sqrt{n}}\right) & \text{for} \quad 0 \leq |t| \leq u, \\[4mm] \Phi(t) - \dfrac{1}{6\sqrt{n}}\phi(t)(\text{sgn}t)\left\{2t^2 + (|t|-u)^2\right\} + o\left(\dfrac{1}{\sqrt{n}}\right) & \text{for} \quad |t| \geq u, \end{cases}$$

where $\Phi(t) = \int_{-\infty}^t \phi(x)dx$ *with* $\phi(x) = (1/\sqrt{2\pi})\, e^{-x^2/2}$.

Proof. Without loss of generality we assume that $\theta = 0$. Putting

$$Z_t = \frac{T_t + tu - \dfrac{|t+u|^3 - |t-u|^3}{12\sqrt{n}}}{u\left(1 - \dfrac{2u}{3\sqrt{n}}\right)^{1/2}},$$

from Lemma 4.4.1, we see that Z_t is asymptotically normally distributed with mean 0 and variance 1. From (4.4.2) we have

$$P_0^n\left\{\sqrt{n}\hat{\theta}_{DL}^u \leq t\right\}$$

$$= P_0^n\{T_t \leq 0\}$$

$$= P_0^n\left\{Z_t \leq \frac{tu - \dfrac{|t+u|^3 - |t-u|^3}{12\sqrt{n}}}{u\left(1 - \dfrac{2u}{3\sqrt{n}}\right)^{1/2}} + o_p\left(\frac{1}{\sqrt{n}}\right)\right\}$$

$$= P_0^n\left\{Z_t \leq \left(t - \frac{|t+u|^3 - |t-u|^3}{12u\sqrt{n}}\right)\left(1 - \frac{2u}{3\sqrt{n}}\right)^{-1/2} + o_p\left(\frac{1}{\sqrt{n}}\right)\right\}$$

$$=P_0^n \left\{ Z_t \leq t - \frac{|t+u|^3 - |t-u|^3}{12u\sqrt{n}} + \frac{tu}{3\sqrt{n}} + o_p\left(\frac{1}{\sqrt{n}}\right) \right\}$$

$$=\Phi(t) - \frac{1}{\sqrt{n}}\phi(t)\left(\frac{|t+u|^3 - |t-u|^3}{12u} - \frac{tu}{3}\right) + o\left(\frac{1}{\sqrt{n}}\right).$$

Hence

$$P_0^n \left\{ \sqrt{n}\hat{\theta}_{DL}^u \leq t \right\}$$

$$= \begin{cases} \Phi(t) - \dfrac{1}{6\sqrt{n}}\phi(t)\dfrac{t(t^2+u^2)}{u} + o\left(\dfrac{1}{\sqrt{n}}\right) & \text{for} \quad 0 \leq |t| \leq u, \\[4mm] \Phi(t) - \dfrac{1}{6\sqrt{n}}\phi(t)(\text{sgn}t)\left\{2t^2 + (|t|-u)^2\right\} + o\left(\dfrac{1}{\sqrt{n}}\right) & \text{for} \quad |t| \geq u, \end{cases}$$

Thus we complete the proof.

From the above theorem we have the following.

Corollary 4.4.1. *The concentration probability of the discretized likelihood estimator* $\hat{\theta}_{DL}^u$ *up to the order* $n^{-1/2}$ *is given by*

$$(4.4.4) \quad P_\theta^n \left\{ \sqrt{n}\left|\hat{\theta}_{DL}^u - \theta\right| \leq t \right\}$$

$$= \begin{cases} 1 - 2\Phi(-t) - \dfrac{1}{3u\sqrt{n}}t(t^2+u^2)\phi(t) + o\left(\dfrac{1}{\sqrt{n}}\right) & \text{for} \quad 0 \leq t \leq u \\[4mm] 1 - 2\Phi(-t) - \dfrac{1}{3\sqrt{n}}\{2t^2 + (t-u)^2\}\phi(t) + o\left(\dfrac{1}{\sqrt{n}}\right) & \text{for} \quad u \leq t. \end{cases}$$

The proof is omitted since it is straightforward from Theorem 4.4.1.

Remark 4.4.1. It is noted by Akahira and Takeuchi (1979, 1981) and Akahira (1986) that the discretized likelihood estimator $\hat{\theta}_{DL}^u$ has the most concentration probability at $t = u$ up to the order $n^{-1/2}$, in the class $\mathbf{A_2}$ of the all second order asymptotically median unbiased estimators, that is, for any $\hat{\theta}$ in $\mathbf{A_2}$ and any θ

$$P_\theta^n \left\{ \sqrt{n}\left|\hat{\theta}_{DL}^u - \theta\right| \leq u \right\} \geq P_\theta^n \left\{ \sqrt{n}\left|\hat{\theta} - \theta\right| \leq u \right\} + o\left(\frac{1}{\sqrt{n}}\right).$$

From Corollary 4.4.1, we have

$$P_\theta^n \left\{ \sqrt{n}\left|\hat{\theta}_{DL}^u - \theta\right| \leq u \right\} = 1 - 2\Phi(-u) - \frac{2u^2}{3\sqrt{n}}\phi(u) + o\left(\frac{1}{\sqrt{n}}\right).$$

4.5. SECOND ORDER ASYMPTOTIC COMPARISON OF THE DISCRETIZED LIKELIHOOD ESTIMATOR WITH OTHERS

In this section we shall compare the discretized likelihood estimator (DLE) with the maximum likelihood estimator (MLE) and some estimator obtained from the order statistics. First, the concentration probability of the maximum likelihood estimator $\hat{\theta}_0$, i.e., the median is given by

$$(4.5.1) \quad P_\theta^n\left\{\sqrt{n}\left|\hat{\theta}_0 - \theta\right| \le t\right\} = 1 - 2\Phi(-t) - \frac{t^2}{\sqrt{n}}\phi(t) + o\left(\frac{1}{\sqrt{n}}\right) \qquad \text{for} \quad t \ge 0,$$

since, from (4.2.11), the asymptotic distribution of $\hat{\theta}_0$ is given by

$$P_\theta^n\left\{\sqrt{n}\left(\hat{\theta}_0 - \theta\right) \le t\right\} = \Phi(t) - \frac{t^2}{2\sqrt{n}}(\operatorname{sgn} t)\phi(t) + o\left(\frac{1}{\sqrt{n}}\right)$$

(see also Akahira and Takeuchi, 1981, page 97). From (4.4.1) and Theorem 4.4.1 it follows that, for any fixed point t, the DLE $\hat{\theta}_{DL}^u$ approximates to the MLE $\hat{\theta}_0$ as $u \to 0$ in the sense that the asymptotic distribution of $\hat{\theta}_{DL}^u$ tends to that of $\hat{\theta}_0$ up to the order $n^{-1/2}$ as $u \to 0$.

Let $X_{(s-k+1)} \le \cdots \le X_{(s+k+1)}$ be the order statistics from a sample of size $n = 2s+1$ from the double exponential distribution, where $k = 0, 1, \ldots, s$. Then we consider $\hat{\theta}_k = \left(X_{(s-k+1)} + X_{(s+k+1)}\right)/2$ as an estimator of θ. If $k = r\sqrt{s} + o(1/\sqrt{s})$ for $r \ge 0$ and $n = 2s + 1$, then the asymptotic distribution of $\hat{\theta}_k$ is given by

$$P_\theta^n\left\{\sqrt{n}\left(\hat{\theta}_k - \theta\right) \le t\right\} = \Phi(t) - \frac{1}{2\sqrt{s}}\phi(t)\left[-rt + \frac{\sqrt{2}}{4}\left\{\left(t - \sqrt{2}r\right)\left|t - \sqrt{2}r\right|\right.\right.$$
$$\left.\left. + \left(t + \sqrt{2}r\right)\left|t + \sqrt{2}r\right|\right\}\right] + o\left(\frac{1}{\sqrt{s}}\right)$$

(see Theorem 4.2.3), hence the concentration probability of $\hat{\theta}_k$ is given by

$$(4.5.2) \quad P_\theta^n\left\{\sqrt{n}\left|\hat{\theta}_k - \theta\right| \le t\right\} = 1 - 2\Phi(-t) - \frac{1}{2\sqrt{s}}\phi(t)\left[-2rt + \frac{\sqrt{2}}{2}\left\{\left(t - \sqrt{2}r\right)\right.\right.$$
$$\left.\left. \cdot \left|t - \sqrt{2}r\right| + \left(t + \sqrt{2}r\right)\left|t + \sqrt{2}r\right|\right\}\right] + o\left(\frac{1}{\sqrt{n}}\right).$$

In comparison of $\hat{\theta}_{DL}^u$, $\hat{\theta}_0$ and $\hat{\theta}_k$ using their concentration probabilities, it seems reasonable to choose $\sqrt{2}r$ as u in the DLE. If $n = 2s+1$ and $u = \sqrt{2}r$, then, from (4.4.4), the concentration probability of the discretized likelihood estimator $\hat{\theta}_{DL}^u$ is given by

$$(4.5.3) \quad P_\theta^n\left\{\sqrt{n}\left|\hat{\theta}_{DL}^{\sqrt{2}r} - \theta\right| \le t\right\}$$
$$= \begin{cases} 1 - 2\Phi(-t) - \dfrac{t(t^2 + 2r^2)}{6r\sqrt{s}}\phi(t) + o\left(\dfrac{1}{\sqrt{s}}\right) & \text{for} \quad 0 \le t \le \sqrt{2}r, \\[2ex] 1 - 2\Phi(-t) - \dfrac{1}{3\sqrt{2}s}(2t^2 + (t - \sqrt{2}r)^2)\phi(t) + o\left(\dfrac{1}{\sqrt{s}}\right) & \text{for} \quad \sqrt{2}r \le t. \end{cases}$$

When $n = 2s + 1$, it follows from (4.5.1) that the concentration probability of the maximum likelihood estimator $\hat{\theta}_0$ is given by

$$(4.5.4) \qquad P_\theta^n \left\{ \sqrt{n} \left| \hat{\theta}_0 - \theta \right| \le t \right\} = 1 - 2\Phi(-t) - \frac{t^2}{\sqrt{2s}} \phi(t) + o\left(\frac{1}{\sqrt{s}} \right).$$

Then we have following theorem.

Theorem 4.5.1. *If $k = r\sqrt{s} + o(1/\sqrt{s})$ for $r \ge 0$ and $n = 2s + 1$, then the order of magnitude of the concentration probability $P_\theta^n\{\sqrt{n}|\hat{\theta} - \theta| \le t\}$ of the estimator $\hat{\theta}$ from the most to the least up to the second order, i.e., the order $n^{-1/2}$, is given by the following table.*

$\hat{\theta}$ t	$\hat{\theta}_{DL}^{\sqrt{2}r}$	$\hat{\theta}_0$	$\hat{\theta}_k$
$0 \le t \le \sqrt{2}(3 - \sqrt{5})r/2$	2	1	3
$\sqrt{2}(3 - \sqrt{5})r/2 \le t \le \sqrt{2}r$	1	2	3
$\sqrt{2}r \le t \le 2\sqrt{2}r$	1	3	2
$2\sqrt{2}r \le t$	2	3	1

Remark 4.5.1. It is clear that the above estimators $\hat{\theta}_{DL}^{\sqrt{2}r}$, $\hat{\theta}_0$ and $\hat{\theta}_k$ are second order asymptotically median unbiased, since the density of the double exponential distribution is symmetric. It is also seen that the three estimators are asymptotically efficient, since their first order asymptotic distribution is given by $\Phi(t)$. From Theorem 4.5.1 it follows that, in the three estimators, the MLE $\hat{\theta}_0$, the DLE $\hat{\theta}_{DL}^{\sqrt{2}r}$ and $\hat{\theta}_k$ have the most concentration probability for $0 \le t \le \sqrt{2}(3 - \sqrt{5})r/2$, $\sqrt{2}(3 - \sqrt{5})r/2 \le t \le 2\sqrt{2}r$ and $2\sqrt{2}r \le t$, respectively. Further it is noted from Remark 4.4.1 that the DLE $\hat{\theta}_{DL}^{\sqrt{2}r}$ has the most concentration probability at $t = \sqrt{2}r$, up to the second order, in the class $\mathbf{A_2}$ which includes the MLE $\hat{\theta}_0$ and $\hat{\theta}_k$. In the case when $n = 2s$, a similar discussion to the above can be also done.

Proof. (i) DLE $\hat{\theta}_{DL}^{\sqrt{2}r}$ and MLE $\hat{\theta}_0$. From (4.4.4) and (4.5.1) it follows that for $0 \le t \le u$

$$P_\theta^n \left\{ \sqrt{n} \left| \hat{\theta}_{DL}^u - \theta \right| \le t \right\} - P_\theta^n \left\{ \sqrt{n} \left| \hat{\theta}_0 - \theta \right| \le t \right\}$$

$$= \frac{\phi(t)}{\sqrt{n}} \left\{ t^2 - \frac{t(t^2 + u^2)}{3u} \right\} + o\left(\frac{1}{\sqrt{n}} \right)$$

$$= -\frac{t\phi(t)}{3u\sqrt{n}}(t^2 - 3tu + u^2) + o\left(\frac{1}{\sqrt{n}} \right).$$

Putting $u = \sqrt{2}r$, we see that

$$\hat{\theta}_{DL}^u \prec \hat{\theta}_0 \qquad \text{for} \quad 0 \le t \le \sqrt{2}(3 - \sqrt{5})r/2,$$
$$\hat{\theta}_{DL}^u \succ \hat{\theta}_0 \qquad \text{for} \quad \sqrt{2}(3 - \sqrt{5})r/2 \le t \le \sqrt{2}r,$$

where $a \prec b$ means that b has more concentration probability than a up to the order $n^{-1/2}$. In a similar way to the above, we have for $u \le t$

$$P_\theta^n \left\{ \sqrt{n} \left| \hat{\theta}_{DL}^u - \theta \right| \le t \right\} - P_\theta^n \left\{ \sqrt{n} \left| \hat{\theta}_0 - \theta \right| \le t \right\}$$

$$= \frac{\phi(t)}{\sqrt{n}} \left\{ t^2 - \frac{2t^2 + (t-u)^2}{3} \right\} + o\left(\frac{1}{\sqrt{n}} \right)$$

$$= \frac{\phi(t)}{3\sqrt{n}} \left\{ t^2 - (t-u)^2 \right\} + o\left(\frac{1}{\sqrt{n}} \right)$$

$$= \frac{\phi(t)u(2t-u)}{3\sqrt{n}} + o\left(\frac{1}{\sqrt{n}} \right)$$

$$\ge \frac{\phi(t)u^2}{3\sqrt{n}} + o\left(\frac{1}{\sqrt{n}} \right),$$

hence, when $u = \sqrt{2}r$, we obtain for $\sqrt{2}r \le t$, $\hat{\theta}_{DL}^{\sqrt{2}r} \succ \hat{\theta}_0$.

(ii) DLE $\hat{\theta}_{DL}^{\sqrt{2}r}$ and $\hat{\theta}_k$. From (4.5.2) and (4.5.3) it follows that for $0 \le t \le \sqrt{2}r$,

$$P_\theta^n \left\{ \sqrt{n} \left| \hat{\theta}_{DL}^{\sqrt{2}r} - \theta \right| \le t \right\} - P_\theta^n \left\{ \sqrt{n} \left| \hat{\theta}_k - \theta \right| \le t \right\}$$

$$= \frac{\phi(t)}{2\sqrt{s}} \left[-2rt + \frac{\sqrt{2}}{2} \left\{ (t - \sqrt{2}r)(\sqrt{2}r - t) + (t + \sqrt{2}r)^2 \right\} - \frac{t(t^2 + 2r^2)}{3r} \right] + o\left(\frac{1}{\sqrt{s}} \right)$$

$$= \frac{t\phi(t)(4r^2 - t^2)}{6r\sqrt{s}} + o\left(\frac{1}{\sqrt{s}} \right)$$

$$\ge \frac{rt\phi(t)}{3\sqrt{s}} + o\left(\frac{1}{\sqrt{s}} \right),$$

hence, for $0 \le t \le \sqrt{2}r$, $\hat{\theta}_{DL}^{\sqrt{2}r} \succ \hat{\theta}_k$. In a similar way to the above, we have for $\sqrt{2}r \le t$,

$$P_\theta^n \left\{ \sqrt{n} \left| \hat{\theta}_{DL}^{\sqrt{2}r} - \theta \right| \le t \right\} - P_\theta^n \left\{ \sqrt{n} \left| \hat{\theta}_k - \theta \right| \le t \right\} = \frac{\phi(t)r(2\sqrt{2}r - t)}{3\sqrt{s}} + o\left(\frac{1}{\sqrt{s}} \right),$$

which yields

$$\hat{\theta}_{DL}^{\sqrt{2}r} \succ \hat{\theta}_k \qquad \text{for} \quad \sqrt{2}r \leq t \leq 2\sqrt{2}r,$$

$$\hat{\theta}_{DL}^{\sqrt{2}r} \prec \hat{\theta}_k \qquad \text{for} \quad 2\sqrt{2}r \leq t.$$

(iii) $\hat{\theta}_0$ and $\hat{\theta}_k$. From (4.5.2) and (4.5.4) it follows that for $0 \leq t \leq \sqrt{2}r$,

$$P_\theta^n \left\{ \sqrt{n} \left| \hat{\theta}_0 - \theta \right| \leq t \right\} - P_\theta^n \left\{ \sqrt{n} \left| \hat{\theta}_k - \theta \right| \leq t \right\}$$

$$= \begin{cases} \sqrt{2}t\phi(t)\dfrac{\sqrt{2}r - t}{2\sqrt{s}} + o\left(\dfrac{1}{\sqrt{s}}\right) & \text{for} \quad 0 \leq t \leq \sqrt{2}r, \\[2ex] -r\phi(t)\dfrac{t - \sqrt{2}r}{\sqrt{s}} + o\left(\dfrac{1}{\sqrt{s}}\right) & \text{for} \quad \sqrt{2}r \leq t, \end{cases}$$

hence $\hat{\theta}_0 \succ \hat{\theta}_k$ for $0 \leq t \leq \sqrt{2}r$, and $\hat{\theta}_0 \prec \hat{\theta}_k$ for $\sqrt{2}r \leq t$. Thus we complete the proof.

In this chapter, the case of the double exponential distribution was treated. As a more general situation, for the case of a continuous density with finite cusps, the first order asymptotic optimality of estimators is discussed by Chernoff and Rubin (1956), Daniels (1961), Williamson (1984) and Prakasa Rao (1968), and the second order asymptotic optimality of estimators is discussed by Akahira (1988a).

CHAPTER 5

ESTIMATION OF A COMMON PARAMETER FOR POOLED SAMPLES FROM THE UNIFORM DISTRIBUTIONS AND THE DOUBLE EXPONENTIAL DISTRIBUTIONS

In this chapter we consider the problem to estimate an unknown real-valued parameter θ based on m samples of size n from the uniform distributions on the interval $(\theta - \xi_i, \theta + \xi_i)$ $(i = 1, \ldots, m)$ with different nuisance parameters which is treated as a typical example in non-regular cases. In some cases the MLE and other estimators will be compared and it will be shown that the MLE based on the pooled sample is not better for both a sample of a fixed size and a large sample. We also consider an estimation problem of a common parameter θ based on m samples of each size n from the double exponential distributions with nuisance parameters τ_i $(i = 1, \ldots, m)$. We obtain the asymptotic expansions of the distributions of some estimators, e.g. the MLE, the weighted median and the weighted mean, and asymptotically compare them up to the second order, i.e. the order $n^{-1/2}$. Further we get the bound for the asymptotic distribution of the all second order asymptotically median unbiased estimators and compare it with their asymptotic distributions up to the order $n^{-1/2}$. Related results can be found in Cohen (1976) and Bhattacharya (1981). For regular cases the reader is referred to Akahira and Takeuchi (1982) and Akahira (1986) where we have samples from m populations with densities $f(x, \theta, \xi_i)$ for $i = 1, \ldots, m$ with the common parameterr θ to be estimated and nuisance parameters ξ_is. It is shown that the MLE is third order asymptotically efficient but there are other estimators which are first and second order asymptotically efficient but have positive deficiency (or loss of information) compared to the MLE. However, in nonregular cases the situation becomes much more complicated and here we deal with two typical cases.

5.1. ESTIMATORS OF A COMMON PARAMETER FOR THE UNIFORM DISTRIBUTIONS

Suppose that it is required to estimate an unknown real-valued parameter θ based on m samples of size n whose values are X_{ij} $(i = 1, \ldots, m; j = 1, \ldots, n)$ from the uniform distributions on the interval $(\theta - \xi_i, \theta + \xi_i)$ with different nuisance parameters

$\xi_i (> 0)$ $(i = 1, \ldots, m)$. For each i let $X_{i(1)} < X_{i(2)} < \cdots < X_{i(n)}$ be order statistics from $X_{i1}, X_{i2}, \ldots, X_{in}$.

An estimator $\hat{\theta}_n$ of θ based on the sample of size n is said to be (asymptotically) median unbiased if

$$P^n_{\theta, \xi_1, \ldots, \xi_m}\left\{\hat{\theta}_n \leq \theta\right\} = P^n_{\theta, \xi_1, \ldots, \xi_m}\left\{\hat{\theta}_n \geq \theta\right\} = \frac{1}{2}$$

$$\left(\lim_{n \to \infty}\left|P^n_{\theta, \xi_1, \ldots, \xi_m}\left\{\hat{\theta}_n \leq \theta\right\} - \frac{1}{2}\right| = \lim_{n \to \infty}\left|P^n_{\theta, \xi_1, \ldots, \xi_m}\left\{\hat{\theta}_n \geq \theta\right\} - \frac{1}{2}\right| = 0 \text{ uniformly in}\right.$$

some neighborhood of θ $\Big)$ (see, e.g. Akahira and Takeuchi, 1981). An estimator $\hat{\theta}_n^*$ is said to be one-sided asymptotically efficient at $(\theta, \xi_1, \ldots, \xi_m)$ in the sense that for any asymptotically median unbiased estimator $\hat{\theta}_n$

$$\lim_{n \to \infty}\left[P^n_{\theta, \xi_1, \ldots, \xi_m}\left\{n\left(\hat{\theta}_n^* - \theta\right) \leq t\right\} - P^n_{\theta, \xi_1, \ldots, \xi_m}\left\{n\left(\hat{\theta}_n - \theta\right) \leq t\right\}\right] \geq 0 \quad \text{for all} \quad t > 0;$$

or

$$\varlimsup_{n \to \infty}\left[P^n_{\theta, \xi_1, \ldots, \xi_m}\left\{n\left(\hat{\theta}_n^* - \theta\right) \leq t\right\} - P^n_{\theta, \xi_1, \ldots, \xi_m}\left\{n\left(\hat{\theta}_n - \theta\right) \leq t\right\}\right] \leq 0 \quad \text{for all} \quad t < 0.$$

We consider some cases.

Case I: $\xi_i = \xi$ $(i = 1, \ldots, m)$ are unknown.

The MLE $\hat{\theta}_{ML}$ of θ based on the pooled sample $\{X_{ij}\}$ is given by

$$\hat{\theta}_{ML} = \frac{1}{2}\left(\min_{1 \leq i \leq m} X_{i(1)} + \max_{1 \leq i \leq m} X_{i(n)}\right).$$

In this case $\hat{\theta}_{ML}$ is one-sided asymptotically efficient at all $(\theta, \xi_1, \ldots, \xi_m)$ (see Akahira and Takeuchi, 1981).

Case II: ξ_i $(i = 1, \ldots, m)$ are known.

The MLE $\hat{\theta}_{ML}^*$ of θ based on the pooled sample $\{X_{ij}\}$ is given by

$$\hat{\theta}_{ML}^* = \frac{1}{2}\left\{\max_{1 \leq i \leq m}\left(X_{i(n)} - \xi_i\right) + \min_{1 \leq i \leq m}\left(X_{i(1)} + \xi_i\right)\right\}.$$

Then it can be shown in a similar way to Akahira (1982a) and Akahira and Takeuchi (1981) that $\hat{\theta}_{ML}^*$ is two-sided asymptotically efficient at $(\theta, \xi_1, \ldots, \xi_m)$ in the sense that for any asymptotically median unbiased estimator $\hat{\theta}_n$

$$\lim_{n \to \infty}\left[P^n_{\theta, \xi_1, \ldots, \xi_m}\left\{n\left|\hat{\theta}_{ML}^* - \theta\right| \leq t\right\} - P^n_{\theta, \xi_1, \ldots, \xi_m}\left\{n\left|\hat{\theta}_n - \theta\right| \leq t\right\}\right] \geq 0 \quad \text{for all} \quad t > 0.$$

(see also Section 7.1, Akahira and Takeuchi (1981, page 72) and Akahira (1982a)).

Case III: ξ_i $(i = 1, \ldots, m)$ *are unknown.*

For each i the MLEs $\hat{\theta}_i$ and $\hat{\xi}_i$ of θ and ξ_i based on the sample $X_{i1}, \ldots, X_{in}$ is given by

$$\hat{\theta}_i = \frac{1}{2}\left(X_{i(1)} + X_{i(n)}\right); \quad \hat{\xi}_i = \frac{1}{2}\left(X_{i(n)} - X_{i(1)}\right),$$

respectively. Let $\hat{\theta}_{ML}$ be the MLE of θ based on the pooled sample $\{X_{ij}\}$. Then it will be shown that $\hat{\theta}_{ML}$ is not two-sided asymptotically efficient at any $(\theta, \xi_1^0, \ldots, \xi_m^0)$. We define

$$\hat{\theta}_n^* = \frac{1}{2}\left\{ \max_{1 \le i \le m}\left(\hat{\theta}_i + \hat{\xi}_i - \xi_i^0\right) + \min_{1 \le i \le m}\left(\hat{\theta}_i - \hat{\xi}_i + \xi_i^0\right) \right\}.$$

Then it can be shown from Case II that $\hat{\theta}_n^*$ is asymptotically locally best estimator of θ at $\xi_i = \xi_i^0$ $(i = 1, \ldots, m)$ in the sense that for any asymptotically median unbiased estimator $\hat{\theta}_n$

$$\lim_{n \to \infty}\left[P_{\theta, \xi_1^0, \ldots, \xi_m^0}^n\left\{ n\left|\hat{\theta}_n^* - \theta\right| \le t \right\} - P_{\theta, \xi_1^0, \ldots, \xi_m^0}^n\left\{ n\left|\hat{\theta}_n - \theta\right| \le t \right\} \right] \ge 0$$

for all $t > 0$.

First we shall obtain the asymptotic formula of the MLE $\hat{\theta}_{ML}$ based on the pooled sample $\{X_{ij}\}$. For each $i = 1, \ldots, m$, let $f_i(x, \theta)$ be a density function of the uniform distribution on the interval $(\theta - \xi_i, \theta + \xi_i)$. Since the likelihood function $L(\theta; \xi_1, \ldots, \xi_m)$ is given by

$$L(\theta; \xi_1, \ldots, \xi_m) = \prod_{i=1}^{m}\prod_{j=1}^{n} f_i(x_j, \theta)$$

$$= \begin{cases} \dfrac{1}{2^n}\left(\dfrac{1}{\prod_{i=1}^{m}\xi_i}\right)^n & \text{for} \quad x_{i(n)} - \xi_i \le \theta \le x_{i(1)} + \xi_i \\[2mm] & \qquad\qquad\qquad (i = 1, \ldots, m); \\[2mm] 0 & \text{otherwise.} \end{cases}$$

In order to obtain the MLE $\hat{\theta}_{ML}$ it is enough to find θ minimizing $\prod_{i=1}^{m}\xi_i$ under the condition $\xi_i \ge \max\left\{\theta - x_{i(1)}, x_{i(n)} - \theta\right\}$ for all i. Let $\hat{\theta}^*$ be some estimator of θ based on the pooled sample $\{X_{ij}\}$. For each i we put $\xi_i^* = \max\left\{\hat{\theta}^* - x_{i(1)}, x_{i(n)} - \hat{\theta}^*\right\}$. Then we have for each i

$$\xi_i^* = \max\left\{\hat{\theta}^* - \hat{\theta}_i + \hat{\xi}_i, \hat{\theta}_i + \hat{\xi}_i - \hat{\theta}^*\right\} = \hat{\xi}_i + \left|\hat{\theta}_i - \hat{\theta}^*\right|.$$

Hence the MLE $\hat{\theta}_{ML}$ is given by $\hat{\theta}^*$ minimizing

$$\prod_{i=1}^{m}\left(\hat{\xi}_i + \left|\hat{\theta}_i - \hat{\theta}^*\right|\right).$$

Since

$$\prod_{i=1}^{m} \left(\hat{\xi}_i + \left| \hat{\theta}_i - \hat{\theta}^* \right| \right) = \prod_{i=1}^{m} \hat{\xi}_i \prod_{i=1}^{m} \left(1 + \frac{\left| \hat{\theta}_i - \hat{\theta}^* \right|}{\hat{\xi}_i} \right),$$

for sufficiently large n it is asymptotically equivalent to

$$\prod_{i=1}^{m} \hat{\xi}_i \left(1 + \sum_{i=1}^{m} \frac{\left| \hat{\theta}_i - \hat{\theta}^* \right|}{\hat{\xi}_i} \right).$$

Hence it is seen that for sufficiently large n the MLE $\hat{\theta}_{ML}$ is asymptotically equivalent to a weighted median by the weights $1/\hat{\xi}_i$ $(i = 1, \ldots, m)$, which is called a quasi-MLE and denoted by $\hat{\theta}_{QML}$.

5.2. COMPARISON OF THE QUASI-MLE, THE WEIGHTED ESTIMATOR AND OTHERS FOR THE UNIFORM DISTRIBUTIONS

In this section we shall discuss the comparison among $\hat{\theta}_{QML}$, the weighted estimator and other estimators. We consider the case when $m = 2$. Then, for each $i = 1, 2$, $X_{i1}, \ldots, X_{in}$ are independently, identically and uniformly distributed random variables on $(\theta - \xi_i, \theta + \xi_i)$. Then, for each $i = 1, 2$, the joint density function $f_n(x, y; \theta, \xi_i)$ of $\hat{\theta}_i$ and $\hat{\xi}_i$ is given by

$$(5.2.1) \quad f_n(x, y; \theta, \xi_i) = \begin{cases} \dfrac{n(n-1)}{2\xi_i^n} y^{n-2} & \text{for} \quad 0 \le y \le \xi_i \quad \text{and} \\ & \qquad \theta - \xi_i + y \le x \le \theta + \xi_i - y; \\ 0 & \text{otherwise.} \end{cases}$$

For each $i = 1, 2$ the density function $f_n(x; \theta, \xi_i)$ of $\hat{\theta}_i$ is given by

$$(5.2.2) \quad f_n(x; \theta, \xi_i) = \begin{cases} \dfrac{n}{2\xi_i^n} \left(\xi_i - |x - \theta| \right)^{n-1} & \text{for} \quad \theta - \xi_i < x < \theta + \xi_i; \\ 0 & \text{otherwise.} \end{cases}$$

Also, for each $i = 1, 2$, the conditional density function $f_n(x|y; \theta, \xi_i)$ of $\hat{\theta}_i$ given $\hat{\xi}_i$ is given by

$$(5.2.3) \quad f_n(x|y; \theta, \xi_i) = \begin{cases} \dfrac{1}{2(\xi_i - y)} & \text{for} \quad \theta - \xi_i + y \le x \le \theta + \xi_i - y; \\ 0 & \text{otherwise.} \end{cases}$$

that is, the conditional distribution of $\hat{\theta}_i$ given $\hat{\xi}_i$ is uniform distribution on the interval $\left(\theta - (\xi_i - \hat{\xi}_i), \theta + (\xi_i - \hat{\xi}_i) \right)$.

For two median unbiased estimators $\hat\theta^1$ and $\hat\theta^2$ of θ, $\hat\theta^1$ is said to be conditionally better than $\hat\theta^2$ given $\hat\xi_1$ and $\hat\xi_2$ if

$$P^n_{\theta,\xi_1,\xi_2}\left\{n\left|\hat\theta^1-\theta\right|\le t\left|\hat\xi_1,\hat\xi_2\right.\right\}\ge P^n_{\theta,\xi_1,\xi_2}\left\{n\left|\hat\theta^2-\theta\right|\le t\left|\hat\xi_1,\hat\xi_2\right.\right\}\qquad\text{for all}\quad t>0,$$

and for simplicity we represent it symbolically as $\hat\theta^1\succ\hat\theta^2|\hat\xi_1,\hat\xi_2$, where $P^n_{\theta,\xi_1,\xi_2}\left\{A|\hat\xi_1,\hat\xi_2\right\}$ denotes the conditional probability of A given $\hat\xi_1$ and $\hat\xi_2$. For two asymptotically median unbiased estimators $\hat\theta^1$ and $\hat\theta^2$ of θ, $\hat\theta^1$ is said to be asymptotically better than $\hat\theta^2$ if

$$\lim_{n\to\infty}\left[P^n_{\theta,\xi_1,\xi_2}\left\{n\left|\hat\theta^1-\theta\right|\le t\right\}-P^n_{\theta,\xi_1,\xi_2}\left\{n\left|\hat\theta^2-\theta\right|\le t\right\}\right]\ge0\quad\text{for all }t>0,$$

and also we denote it symbolically by $\hat\theta^1\underset{\text{as.}}{\succ}\hat\theta^2$.

Theorem 5.2.1. In case $\xi_1=\xi_2$, then for $\hat\xi_1<\hat\xi_2$ given

$$\frac{\hat\xi_2\hat\theta_1+\hat\xi_1\hat\theta_2}{\hat\xi_1+\hat\xi_2}\succ\frac{\hat\xi_2^2\hat\theta_1+\hat\xi_1^2\hat\theta_2}{\hat\xi_1^2+\hat\xi_2^2}\succ\hat\theta_{QML}=\hat\theta_1\left|\hat\xi_1,\hat\xi_2\right..$$

In case $\xi_1=\xi_2$, then for $\hat\xi_1>\hat\xi_2$ given

$$\frac{\hat\xi_2\hat\theta_1+\hat\xi_1\hat\theta_2}{\hat\xi_1+\hat\xi_2}\succ\frac{\hat\xi_2^2\hat\theta_1+\hat\xi_1^2\hat\theta_2}{\hat\xi_1^2+\hat\xi_2^2}\succ\hat\theta_{QML}=\hat\theta_2\left|\hat\xi_1,\hat\xi_2\right..$$

Proof. From (5.2.3) we see that the conditional density of $\hat\theta_1-\theta$ and $\hat\theta_2-\theta$ given $\hat\xi_1$ and $\hat\xi_2$ is given by

$$f_n\left(x_1,x_2|\hat\xi_1,\hat\xi_2\right)=\begin{cases}\dfrac{1}{4\tau_1\tau_2}&\text{for}\quad|x_1|<\tau_1\quad\text{and}\quad|x_2|<\tau_2;\\[2mm]0&\text{otherwise,}\end{cases}$$

where $\tau_i=\xi_i-\hat\xi_i$ $(i=1,2)$. Then the conditional density function $f_n(y|\hat\xi_1,\hat\xi_2)$ of $\hat\theta_0-\theta$ with a linear estimator $\hat\theta_0=c_1\hat\theta_1+c_2\hat\theta_2$ given $\hat\xi_1$ and $\hat\xi_2$ is given by

$$(5.2.4)\quad f_n\left(y|\hat\xi_1,\hat\xi_2\right)=\begin{cases}\dfrac{1}{4c_1c_2\tau_1\tau_2}(c_1\tau_1+c_2\tau_2+y)&\text{for}\quad y<-c_1\tau_1+c_2\tau_2;\\[2mm]\dfrac{1}{2c_1\tau_1}&\text{for}\quad|y|<c_1\tau_1-c_2\tau_2;\\[2mm]\dfrac{1}{4c_1c_2\tau_1\tau_2}(c_1\tau_1+c_2\tau_2-y)&\text{for}\quad y>c_1\tau_1-c_2\tau_2\end{cases}$$

for $c_1\tau_1 > c_2\tau_2$. For $c_1\tau_1 < c_2\tau_2$, the conditional density function is also given by

$$(5.2.5) \quad f_n\left(y|\hat{\xi}_1, \hat{\xi}_2\right) = \begin{cases} \dfrac{1}{4c_1c_2\tau_1\tau_2}(c_1\tau_1 + c_2\tau_2 - y) & \text{for } y > -c_1\tau_1 + c_2\tau_2; \\[2ex] \dfrac{1}{2c_2\tau_2} & \text{for } |y| < -c_1\tau_1 + c_2\tau_2; \\[2ex] \dfrac{1}{4c_1c_2\tau_1\tau_2}(c_1\tau_1 + c_2\tau_2 + y) & \text{for } y < c_1\tau_1 - c_2\tau_2. \end{cases}$$

If

$$c_i = \hat{\xi}_j / \left(\hat{\xi}_1 + \hat{\xi}_2\right) = c_i' \quad \text{(say)} \qquad (i \neq j; \ i, j = 1, 2),$$

then

$$(5.2.6) \qquad \frac{1}{c_1'\tau_1} = \frac{\hat{\xi}_1 + \hat{\xi}_2}{\hat{\xi}_2\left(\hat{\xi}_1 - \hat{\xi}_1\right)}; \qquad \frac{1}{c_2'\tau_2} = \frac{\hat{\xi}_1 + \hat{\xi}_2}{\hat{\xi}_1\left(\hat{\xi}_2 - \hat{\xi}_2\right)}.$$

If

$$c_i = \hat{\xi}_j^2 / \left(\hat{\xi}_1^2 + \hat{\xi}_2^2\right) = c_i'' \quad \text{(say)} \qquad (i \neq j; \ i, j = 1, 2),$$

then

$$(5.2.7) \qquad \frac{1}{c_1''\tau_1} = \frac{\hat{\xi}_1^2 + \hat{\xi}_2^2}{\hat{\xi}_2^2\left(\hat{\xi}_1 - \hat{\xi}_1\right)}; \qquad \frac{1}{c_2''\tau_2} = \frac{\hat{\xi}_1^2 + \hat{\xi}_2^2}{\hat{\xi}_1^2\left(\hat{\xi}_2 - \hat{\xi}_2\right)}.$$

Hence we have the following :

$$(5.2.8) \quad \hat{\xi}_2 \gtreqless \hat{\xi}_1 \ \text{ if and only if } \ \frac{1}{c_1'\tau_1} = \frac{\hat{\xi}_1 + \hat{\xi}_2}{\hat{\xi}_2\left(\hat{\xi}_1 - \hat{\xi}_1\right)} \gtreqless \frac{\hat{\xi}_1^2 + \hat{\xi}_2^2}{\hat{\xi}_2^2\left(\hat{\xi}_1 - \hat{\xi}_1\right)} = \frac{1}{c_1''\tau_1};$$

$$(5.2.9) \quad \hat{\xi}_2 \gtreqless \hat{\xi}_1 \ \text{ if and only if } \ \frac{1}{c_2'\tau_2} = \frac{\hat{\xi}_1 + \hat{\xi}_2}{\hat{\xi}_1\left(\hat{\xi}_2 - \hat{\xi}_2\right)} \lesseqgtr \frac{\hat{\xi}_1^2 + \hat{\xi}_2^2}{\hat{\xi}_1^2\left(\hat{\xi}_2 - \hat{\xi}_2\right)} = \frac{1}{c_2''\tau_2}.$$

On the other hand as is seen from the above discussion on the quasi-MLE $\hat{\theta}_{QML}$ based on the pooled sample $\{X_{ij}\}$ is given by

$$\hat{\theta}_{QML} = \begin{cases} \hat{\theta}_1 & \text{if } \hat{\xi}_1 < \hat{\xi}_2; \\ \hat{\theta}_2 & \text{if } \hat{\xi}_1 > \hat{\xi}_2. \end{cases}$$

The conditional density of the quasi-MLE $\hat{\theta}_{QML}$ given $\hat{\xi}_i$ is given by (5.2.3). Let $\xi_1 = \xi_2 = \xi$. In the case when $c_i = c_i' = \hat{\xi}_j / \left(\hat{\xi}_1 + \hat{\xi}_2\right)$ $(i \neq j; i, j = 1, 2)$,

$$\hat{\xi}_1 \lesseqgtr \hat{\xi}_2 \ \text{ if and only if } \ \hat{\xi}_1\left(\xi - \hat{\xi}_2\right) \lesseqgtr \hat{\xi}_2\left(\xi - \hat{\xi}_1\right), \qquad \text{i.e., } \ c_2\tau_2 \lesseqgtr c_1\tau_1.$$

Hence we have

(5.2.10) $\qquad \hat{\xi}_1 \leq \hat{\xi}_2$ if and only if $c_1' \tau_1 + c_2' \tau_2 < \begin{cases} \tau_1 \\ \tau_2 \end{cases}$.

We also obtain

$$c_1'' \tau_1 + c_2'' \tau_2 - (c_1' \tau_1 + c_2' \tau_2) = \frac{\hat{\xi}_1 \hat{\xi}_2 \left(\hat{\xi}_1 - \hat{\xi}_2\right)^2}{\left(\hat{\xi}_1 + \hat{\xi}_2\right)\left(\hat{\xi}_1^2 + \hat{\xi}_2^2\right)} \geq 0.$$

From (5.2.4) to (5.2.10) the conclusion follows as required.

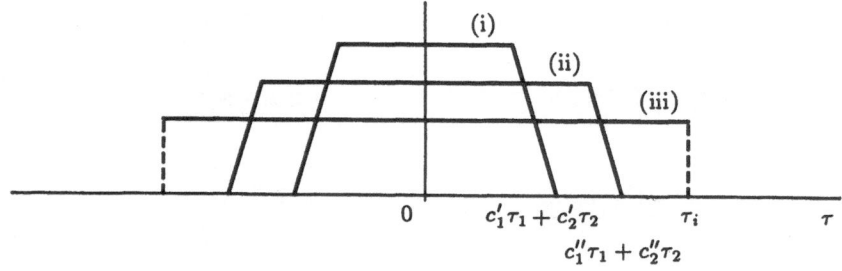

Figure 5.2.1. Comparison of the conditional densities of $c_1 \hat{\theta} + c_2 \hat{\theta}_2 - \theta$ given $\hat{\xi}_1$ and $\hat{\xi}_2$ with (i) $c_i = c_i' = \hat{\xi}_j \big/ \left(\hat{\xi}_1 + \hat{\xi}_2\right)$ $(i \neq j; i,j = 1, 2)$ and (ii) $c_i = c_i'' = \hat{\xi}_j^2 \big/ \left(\hat{\xi}_1^2 + \hat{\xi}_2^2\right)$ $(i \neq j; i,j = 1, 2)$ and of (iii) the quasi-MLE $\hat{\theta}_{QML}$ given $\hat{\xi}_i$. They are given by (5.2.5) and (5.2.3), respectively.

Theorem 5.2.2. *If* $\xi_1 = \xi_2 = \xi$, *then*

$$\frac{1}{2} \left(\hat{\theta}_1 + \hat{\theta}_2\right) \underset{\text{a.s.}}{\succ} \hat{\theta}_{QML} = \begin{cases} \hat{\theta}_1 & \text{if } \hat{\xi}_1 < \hat{\xi}_2; \\ \hat{\theta}_2 & \text{if } \hat{\xi}_1 > \hat{\xi}_2. \end{cases}$$

Proof. By (5.2.2) it follows that for each $i = 1, 2$ the asymptotic density of $n\left(\hat{\theta}_i - \theta\right)$ is given by

(5.2.11) $\qquad\qquad f_i(x) = \frac{1}{2\xi_i} e^{-|x|/\xi_i}.$

Then the characteristic function $\phi(t)$ of the asymptotic density function of $n\left[\left\{(\hat{\theta}_1 + \hat{\theta}_2)/2\right\} - \theta\right]$ is given by

$$\phi(t) = \frac{1}{\left(1 + \xi^2 t^2/4\right)^2}.$$

When $\xi_1 = \xi_2 = \xi$, $\hat{\theta}_1$ and $\hat{\theta}_2$ have the same asymptotic density as (5.2.11) with $\xi_i = \xi$ $(i = 1, 2)$. We may represent $\phi(t)$ as follows :

(5.2.12)
$$\phi(t) = \frac{1 - \xi^2 t^2/4}{2(1 + \xi^2 t^2/4)^2} + \frac{1}{2(1 + \xi^2 t^2/4)}.$$

Since

$$\int_{-\infty}^{\infty} \frac{1}{\xi} \exp\left(-\frac{2}{\xi}|y|\right) \exp(ity)dy = \frac{1}{2} \int_{-\infty}^{\infty} \exp(-|x|) \exp\left(i\frac{\xi t}{2}x\right) dx$$
$$= \frac{1}{1 + \xi^2 t^2/4},$$

$$\int_{-\infty}^{\infty} \frac{2}{\xi^2}|y| \exp\left(-\frac{2}{\xi}|y|\right) \exp(ity)dy = \frac{1}{2} \int_{-\infty}^{\infty} |x| \exp(-|x|) \exp\left(i\frac{\xi t}{2}x\right) dx$$
$$= \frac{1 - \xi^2 t^2/4}{\left(1 + \xi^2 t^2/4\right)^2},$$

it follows from (5.2.12) that the asymptotic density of $n\left[\{(\hat{\theta}_1 + \hat{\theta}_2)/2\} - \theta\right]$ is given by

(5.2.13)
$$f(x) = \frac{1}{2\xi}\left(1 + 2\frac{|x|}{\xi}\right) \exp\left(-\frac{2}{\xi}|x|\right).$$

From (5.2.11) and (5.2.13) the conclusion follows as required.

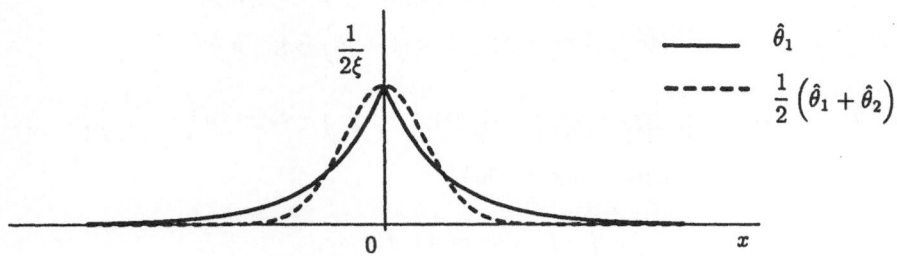

Figure 5.2.2. Comparison of the asymptotic densities of $n(\hat{\theta}_1 - \theta)$ and $n\left[\{(\hat{\theta}_1 + \hat{\theta}_2)/2\} - \theta\right]$ given by (5.2.11) and (5.2.13), respectively.

Theorem 5.2.3. *Suppose that $\xi_1 < \xi_2$ and $c_1\xi_1 \neq c_2\xi_2$, where c_1 and c_2 are constants. Let $\hat{\theta}_0 = c_1\hat{\theta}_1 + c_2\hat{\theta}_2$. Then*
$$\hat{\theta}_{QML} = \hat{\theta}_1 \underset{as.}{\succ} \hat{\theta}_0$$
in some neighborhood of θ in the sense that
$$\lim_{n \to \infty} \left[P_{\theta,\xi_1,\xi_2}^n\left\{n\left|\hat{\theta}_{QML} - \theta\right| \leq t\right\} - P_{\theta,\xi_1,\xi_2}^n\left\{n\left|\hat{\theta}_0 - \theta\right| \leq t\right\}\right] \geq 0 \qquad \text{for all} \quad t \leq t_0.$$

Estimation of a common parameter

Further

$$\hat{\theta}_0 \underset{as.}{\succ} \hat{\theta}_{QML} = \hat{\theta}_1$$

in far away from θ *in the sense that*

$$\lim_{n \to \infty} \left[P^n_{\theta,\xi_1,\xi_2} \left\{ n \left| \hat{\theta}_0 - \theta \right| \leq t \right\} - P^n_{\theta,\xi_1,\xi_2} \left\{ n \left| \hat{\theta}_{QML} - \theta \right| \leq t \right\} \right] \geq 0 \qquad \text{for all} \quad t > t_0.$$

Proof. By (5.2.2) it follows that, for each $i = 1, 2$, the asymptotic density of $n(\hat{\theta}_i - \theta)$ is given by

$$(5.2.14) \qquad\qquad f_i(x) = \frac{1}{2\xi_i} \exp(-|x|/\xi_i),$$

and also its characteristic function $\phi_i(t)$ is given by

$$\phi_i(t) = \frac{1}{1 + \xi_i^2 t^2}.$$

Hence the characteristic function $\phi^*(t)$ of the asymptotic density of $n(\hat{\theta}_0 - \theta) = n(c_1 \hat{\theta}_1 + c_2 \hat{\theta}_2 - \theta)$ is given by

$$\phi^*(t) = \frac{1}{(1 + c_1^2 \xi_1^2 t^2)(1 + c_2^2 \xi_2^2 t^2)}.$$

Since

$$\phi^*(t) = \frac{1}{(c_1^2 \xi_1^2 - c_2^2 \xi_2^2)} \left(\frac{c_1^2 \xi_1^2}{1 + c_1^2 \xi_1^2 t^2} - \frac{c_2^2 \xi_2^2}{1 + c_2^2 \xi_2^2 t^2} \right),$$

it follows that the asymptotic density $f(x)$ of $n(\hat{\theta}_0 - \theta)$ is given by

$$(5.2.15) \qquad f(x) = \frac{1}{2(c_1^2 \xi_1^2 - c_2^2 \xi_2^2)} \left\{ c_1 \xi_1 \exp\left(-\frac{|x|}{c_1 \xi_1} \right) - c_2 \xi_2 \exp\left(-\frac{|x|}{c_2 \xi_2} \right) \right\}.$$

There exists a positive number t_0 such that

$$\int_0^{t_0} f_1(x) dx = \int_0^{t_0} f(x) dx,$$

where $f_1(x)$ and $f(x)$ are given by (5.2.14) and (5.2.15), respectively. Hence we complete the proof.

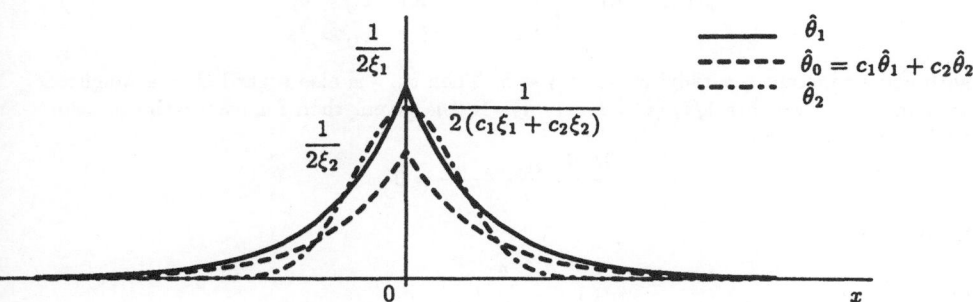

Figure 5.2.3. Comparison of the asymptotic densities of $n(\hat{\theta}_1 - \theta)$, $n(\hat{\theta}_0 - \theta) = n(c_1\hat{\theta}_1 + c_2\hat{\theta}_2 - \theta)$ and $n(\hat{\theta}_2 - \theta)$ given by (5.2.14), (5.2.15) and (5.2.14), respectively, when $\xi_1 < \xi_2$.

Remark 5.2.1. From (5.2.14) and (5.2.15) it is easily seen that

$$\hat{\theta}_0 \underset{\text{as.}}{\succ} \hat{\theta}_2 \quad \text{and} \quad \hat{\theta}_{QML} = \hat{\theta}_1 \underset{\text{as.}}{\succ} \hat{\theta}_2.$$

The case $\xi_1 > \xi_2$ may be treated quite similarly.

5.3. ESTIMATORS OF A COMMON PARAMETER FOR THE DOUBLE EXPONENTIAL DISTRIBUTIONS

Let X_{ij} $(i = 1, \ldots, m; j = 1, \ldots, n)$ be m sets of independent samples each of size n. Suppose that for each i, X_{ij} $(j = 1, \ldots, n)$ have the following density function

$$f(x, \theta, \tau_i) = \frac{1}{2\tau_i} \exp\left(-\frac{|x - \theta|}{\tau_i}\right) \qquad \text{for} \quad -\infty < x < \infty,$$

where θ and τ_i are real and positive valued parameters, respectively.

We denote the log-likelihood function for the all m samples by $L(\theta, \tau_1, \ldots, \tau_m)$. Then we have

$$(5.3.1) \qquad L(\theta, \tau_1, \ldots, \tau_m) = -\sum_{i=1}^{m} \frac{1}{\tau_i} \sum_{j=1}^{n} |x_{ij} - \theta| - n \sum_{i=1}^{m} \log \tau_i - mn \log 2.$$

If τ_i $(i = 1, \ldots, m)$ are known, the maximum likelihood estimator $\hat{\theta}_{ML}^0$ is given as the solution of the equation

$$(5.3.2) \qquad \sum_{i=1}^{m} \frac{1}{\tau_i} \sum_{j=1}^{n} \text{sgn}\,(x_{ij} - \theta) = 0,$$

where

$$\text{sgn}\,(x_{ij} - \theta) = \begin{cases} -1 & \text{for} \quad x_{ij} < \theta; \\ \gamma & \text{for} \quad x_{ij} = \theta; \\ 1 & \text{for} \quad x_{ij} > \theta; \end{cases}$$

with some constant γ satisfying $-1 < \gamma < 1$. Then $\hat{\theta}^0_{ML}$ is also regarded as a weighted median by the weights $1/\tau_i$ $(i = 1, \ldots, m)$. If θ is given, then for each i the solution $\hat{\tau}_i(\theta)$ of the equation

$$\frac{\partial L(\theta, \tau_1, \ldots, \tau_m)}{\partial \tau_i} = 0$$

is given by

(5.3.3) $$\hat{\tau}_i(\theta) = \frac{1}{n} \sum_{j=1}^{n} |x_{ij} - \theta|.$$

Substituting (5.2.3) in (5.2.1) we have

(5.3.4) $$L\left(\theta, \hat{\tau}_1(\theta), \ldots, \hat{\tau}_m(\theta)\right) = mn \left(\log \frac{n}{2} - 1\right) - n \sum_{i=1}^{m} \log \sum_{j=1}^{n} |x_{ij} - \theta|.$$

Since in order to get θ maximizing (5.3.1), it is enough to obtain θ minimizing $\sum_{i=1}^{m} \log \sum_{j=1}^{n} |x_{ij} - \theta|$ by (5.3.4), and such a θ is given as a solution $\hat{\theta}_{WM}$ of the equation

(5.3.5) $$\sum_{i=1}^{m} \frac{\sum_{j=1}^{n} \text{sgn}\,(x_{ij} - \theta)}{\sum_{j=1}^{n} |x_{ij} - \theta|} = 0.$$

It is seen by (5.3.3) and (5.3.5) that the estimator $\hat{\theta}_{WM}$ is a weighted median by the weights $1/\hat{\tau}_i\left(\hat{\theta}_{WM}\right)$ $(i = 1, \ldots, m)$.

We also consider other estimators. If τ_i $(i = 1, \ldots, m)$ are known and unknown, we have the weighted means $\hat{\theta}^0$ and $\hat{\theta}^*$ by the weights $1/\tau_i^2$ $(i = 1, \ldots, m)$ and $1/\hat{\tau}_i^2\left(\hat{\theta}_i\right)$ $(i = 1, \ldots, m)$, respectively, i.e.

$$\hat{\theta}^0 = \sum_{i=1}^{m} \frac{1}{\tau_i^2} \hat{\theta}_i \bigg/ \sum_{i=1}^{m} \frac{1}{\tau_i^2}\,; \qquad \hat{\theta}^* = \sum_{i=1}^{m} \frac{1}{\hat{\tau}_i^2} \hat{\theta}_i \bigg/ \sum_{i=1}^{m} \frac{1}{\hat{\tau}_i^2}\,,$$

where $\hat{\tau}_i = \hat{\tau}_i\left(\hat{\theta}_i\right)$ with $\hat{\theta}_i = \text{med}_{1 \le j \le n} X_{ij}$ $(i = 1, \ldots, m)$.

In the next section we shall compare the above asymptotically efficient estimators up to the second order i.e. the order $n^{-1/2}$ and also obtain the bound for the asymptotic distribution of second order asymptotically median unbiased estimators up to the order $n^{-1/2}$.

5.4. SECOND ORDER ASYMPTOTIC COMPARISON OF THE ESTIMATORS FOR THE DOUBLE EXPONENTIAL DISTRIBUTIONS

First we shall obtain the asymptotic expansion of the distribution of the MLE $\hat{\theta}^0_{ML}$ up to the order $n^{-1/2}$ in the case when τ_i $(i = 1, \ldots, m)$ are known. Since $\hat{\theta}^0_{ML}$ is the solution of the equation (5.3.2), the function of θ given by the left-hand side of (5.3.2) is locally monotone decreasing in a neighborhood of $\hat{\theta}^0_{ML}$. Hence it follows that for any positive (negative) c, $\hat{\theta}^0_{ML} \lessgtr \theta + (c/\sqrt{n})$ if and only if

$$\sum_{i=1}^{m} \frac{1}{\tau_i} \sum_{j=1}^{n} \operatorname{sgn}\left(x_{ij} - \theta - \frac{c}{\sqrt{n}}\right) \lessgtr 0.$$

Then we have the following.

Theorem 5.4.1. *The bound for the asymptotic distribution of second order AMU estimators and the asymptotic expansion of the distribution of the estimators $\hat{\theta}^0_{ML}$ and $\hat{\theta}_{WM}$ up to the order $n^{-1/2}$ have the form of*

$$\Phi(t) - \frac{\alpha p_3 t^2}{p_2^{3/2} \sqrt{n}} \phi(t) \operatorname{sgn} t + o\left(\frac{1}{\sqrt{n}}\right)$$

with α given by the following :

Estimator	α
The bound	1/6
$\hat{\theta}^0_{ML}$	1/2
$\hat{\theta}_{WM}$	1/2

Table 5.4.1.

where $p_k = \sum_{i=1}^{m} 1/\tau_i^k$ $(k = 2, 3)$.

Remark 5.4.1. By Theorem 5.4.1 we see that both of the estimators $\hat{\theta}^0_{ML}$ and $\hat{\theta}_{WM}$ are asymptotically equivalent up to the order $n^{-1/2}$ in spite that $\hat{\theta}^0_{ML}$ and $\hat{\theta}_{WM}$ are the estimators for known τ_i $(i = 1, \ldots, m)$ and unknown τ_i's, respectively, but they are not second order asymptotically efficient in the sense that their asymptotic distributions do not attain the bound uniformly.

Proof. In order to obtain the asymptotic expansion of the distribution of the MLE $\hat{\theta}^0_{ML}$, without loss of generality we assume that $\theta = 0$. We put $W_{ij} = \operatorname{sgn}\left(X_{ij} - cn^{-1/2}\right)$ $(i = 1, \ldots, m; j = 1, \ldots, n)$. Then we have for any fixed real number c

(5.4.1) $$E_0[W_{ij}] = -\frac{c}{\tau_i \sqrt{n}} + \frac{c^2 \operatorname{sgn} c}{2\tau_i^2 n} + o\left(\frac{1}{n}\right).$$

Putting

$$\psi_i(a) = \frac{1}{\tau_i} \sum_{j=1}^{n} \operatorname{sgn}\left(X_{ij} - a\right) \qquad (i = 1, \ldots, m),$$

we obtain by (5.4.1)

$$(5.4.2) \qquad E_0\left[\frac{1}{\sqrt{n}}\sum_{i=1}^m \psi_i\left(\frac{c}{\sqrt{n}}\right)\right] = \frac{1}{\sqrt{n}}\sum_{i=1}^m \frac{1}{\tau_i}\sum_{j=1}^n E_0\left(W_{ij}\right)$$

$$= -cp_2 + \frac{p_3 c^2 \mathrm{sgn} c}{2\sqrt{n}} + o\left(\frac{1}{\sqrt{n}}\right),$$

where $p_k = \sum_{i=1}^m 1/\tau_i^k$ $(k=2,3)$. Since

$$E_0\left[W_{ij}^2\right] = E_0\left[\left\{\mathrm{sgn}\left(X_{ij} - \frac{c}{\sqrt{n}}\right)\right\}^2\right] = 1,$$

it follows from (5.4.1) that the asymptotic variance of W_{ij} at $\theta=0$ is given by

$$(5.4.3) \qquad V_0\left(W_{ij}\right) = 1 + \frac{c^2}{\tau_i^2 n} + o\left(\frac{1}{n}\right).$$

Then we obtain from (5.4.3)

$$(5.4.4) \qquad V_0\left(\frac{1}{\sqrt{n}}\sum_{i=1}^m \psi_i\left(\frac{c}{\sqrt{n}}\right)\right) = p_2 + \frac{p_4 c^2}{n} + o\left(\frac{1}{n}\right).$$

We also have

$$(5.4.5) \qquad E_0\left[\left\{\psi_i\left(\frac{c}{\sqrt{n}}\right)\right\}^2\right] = \frac{n}{\tau_i^2}\left\{1 + (n-1)\mu_i^2\right\};$$

$$(5.4.6) \qquad E_0\left[\left\{\psi_i\left(\frac{c}{\sqrt{n}}\right)\right\}^3\right] = \frac{n}{\tau_i^3}\left\{\mu_i + 3(n-1)\mu_i + (n-1)(n-2)\mu_i^3\right\},$$

where

$$\mu_i = E_0\left(W_{ij}\right) = -\frac{c}{\tau_i\sqrt{n}} + \frac{c^2 \mathrm{sgn} c}{2\tau_i^2 n} + o\left(\frac{1}{n}\right).$$

By (5.4.5) and (5.4.6) we obtain the third order cumulant

$$(5.4.7) \qquad K_{3,0}\left(\frac{1}{\sqrt{n}}\sum_{i=1}^m \psi_i\left(\frac{c}{\sqrt{n}}\right)\right) = \frac{2c}{n}p_4 + o\left(\frac{1}{n}\right).$$

Then it follows from (5.4.2), (5.4.4) and (5.4.7) that the asymptotic expansion of the distribution of the MLE $\hat{\theta}_{ML}^0$ up to the order $n^{-1/2}$ is given by

$$(5.4.8)\, P_{0,\tau_1,\ldots,\tau_m}^n\left\{\sqrt{n}\hat{\theta}_{ML}^0 \le c\right\} = P_{0,\tau_1,\ldots,\tau_m}^n\left\{\frac{1}{\sqrt{n}}\sum_{i=1}^m \frac{1}{\tau_i}\sum_{j=1}^n \mathrm{sgn}\left(X_{ij} - \frac{c}{\sqrt{n}}\right) \le 0\right\}$$

$$= \Phi\left(c\sqrt{\bar{p}_2}\right) - \frac{p_3 c^2}{2\sqrt{p_2 n}}\phi\left(c\sqrt{\bar{p}_2}\right)\mathrm{sgn} c + o\left(\frac{1}{\sqrt{n}}\right),$$

where $\Phi(x) = \int_{-\infty}^{x} \phi(u)du$ with $\phi(u) = (1/\sqrt{2\pi})e^{-u^2/2}$. We also have by (5.4.8)

$$(5.4.9) \quad P_{\theta,\tau_1,\ldots,\tau_m}^{n}\left\{\sqrt{p_2 n}\left(\hat{\theta}_{ML}^{0} - \theta\right) \leq t\right\} = \Phi(t) - \frac{p_3 t^2}{2p_2^{3/2}\sqrt{n}}\phi(t)\text{sgn}t + o\left(\frac{1}{\sqrt{n}}\right).$$

Next we shall obtain the asymptotic expansion of the distribution of the estimator $\hat{\theta}_{WM}$, i.e. the weighted median by the weights $1\big/\hat{\tau}_i\left(\hat{\theta}_{WM}\right)$ $(i = 1,\ldots,m)$ up to the order $n^{-1/2}$ in the case when τ_i $(i = 1,\ldots,m)$ are unknown. Since $\hat{\theta}_{WM}$ is the solution of the equation (5.3.5), the function of θ given by the left-hand side of (5.3.5) is locally monotone decreasing in a neighborhood of $\hat{\theta}_{WM}$. Hence it follows that for any fixed positive (negative) c, $\hat{\theta}_{WM} \lesseqgtr \theta + (c/\sqrt{n})$ if and only if

$$\sum_{i=1}^{m} \frac{\dfrac{1}{\sqrt{n}}\displaystyle\sum_{j=1}^{n}\text{sgn}\left(x_{ij} - \theta - \dfrac{c}{\sqrt{n}}\right)}{\dfrac{1}{n}\displaystyle\sum_{j=1}^{n}\left|x_{ij} - \theta - \dfrac{c}{\sqrt{n}}\right|} \lesseqgtr 0.$$

In order to obtain the asymptotic expansion of the distribution of the estimator $\hat{\theta}_{WM}$, without loss of generality we assume that $\theta = 0$. For each $i = 1,\ldots,m$ we put

$$Y_i = \frac{1}{\sqrt{n}}\sum_{j=1}^{n}\text{sgn}\left(X_{ij} - \frac{c}{\sqrt{n}}\right); \qquad Z_i = \frac{1}{\sqrt{n}}\sum_{j=1}^{n}\left|X_{ij} - \frac{c}{\sqrt{n}}\right| - \tau_i\sqrt{n}.$$

Then we have

$$(5.4.10) \quad \sum_{i=1}^{m} \frac{\dfrac{1}{\sqrt{n}}\displaystyle\sum_{j=1}^{n}\text{sgn}\left(X_{ij} - \dfrac{c}{\sqrt{n}}\right)}{\dfrac{1}{n}\displaystyle\sum_{j=1}^{n}\left|X_{ij} - \dfrac{c}{\sqrt{n}}\right|}$$

$$= \sum_{i=1}^{m} Y_i \frac{1}{\tau_i}\left\{1 - \frac{Z_i}{\tau_i\sqrt{n}} + \frac{Z_i^2}{\tau_i^2 n} + o_p\left(\frac{1}{n}\right)\right\} = \sum_{i=1}^{m} Y_i U_i \qquad \text{(say)}.$$

Since

$$E_0\left[\text{sgn}\left(X_{ij} - \frac{c}{\sqrt{n}}\right)\left|X_{ij} - \frac{c}{\sqrt{n}}\right|\right] = -\frac{c}{\sqrt{n}} \qquad (i = 1,\ldots,m; j = 1,\ldots,n)$$

and

$$(5.4.11) \quad E_0\left[\left|X_{ij} - \frac{c}{\sqrt{n}}\right|\right] = \tau_i + \frac{c^2}{2\tau_i n} + o\left(\frac{1}{n}\right) \qquad (i = 1,\ldots,m; j = 1,\ldots,n),$$

it follows that

$$(5.4.12) \qquad\qquad\qquad E_0\left(Y_i Z_i\right) = o(1).$$

Since, for each $i = 1, \ldots, m$, Y_i and Z_i are asymptotically normally distributed, it follows from (5.4.11) that Y_i and Z_i are asymptotically independent. By (5.4.1) and (5.4.11) we have for each $i = 1, \ldots, m$

(5.4.13) $$E_0[Y_i] = -\frac{c}{\tau_i} + \frac{c^2 \text{sgn} c}{2\tau_i^2 \sqrt{n}} + o\left(\frac{1}{\sqrt{n}}\right);$$

(5.4.14) $$E_0[Z_i] = \frac{c^2}{2\tau_i \sqrt{n}} + o\left(\frac{1}{\sqrt{n}}\right).$$

By (5.4.13) and (5.4.14) we have

(5.4.15) $$E_0[Y_i U_i] = -\frac{c}{\tau_i^2} + \frac{c^2 \text{sgn} c}{2\tau_i^3 \sqrt{n}} + o\left(\frac{1}{\sqrt{n}}\right).$$

Hence we obtain

(5.4.16) $$E_0\left[\sum_{i=1}^{m} Y_i U_i\right] = -p_2 c + \frac{p_3 c^2 \text{sgn} c}{2\sqrt{n}} + o\left(\frac{1}{\sqrt{n}}\right).$$

By (5.4.1) we obtain for each $i = 1, \ldots, m$

(5.4.17) $$E_0[Y_i^2] = 1 + \frac{c^2}{\tau_i^2} - \frac{c^3 \text{sgn} c}{\tau_i^3 \sqrt{n}} + o\left(\frac{1}{\sqrt{n}}\right).$$

Since

$$E_0\left[\left(X_{ij} - \frac{c}{\sqrt{n}}\right)^2\right] = 2\tau_i^2 + \frac{c^2}{n},$$

it follows from (5.4.11) that

(5.4.18) $$E_0[Z_i^2] = \tau_i^2 + o(1).$$

Since by (5.4.17) and (5.4.18)

(5.4.19) $$E_0[Y_i^2 U_i^2] = \frac{1}{\tau_i^2} + \frac{c^2}{\tau_i^4} - \frac{c^3 \text{sgn} c}{\tau_i^5 \sqrt{n}} + o\left(\frac{1}{n}\right),$$

it follows from (5.4.16) that

$$V_0(Y_i U_i) = \frac{1}{\tau_i^2} + o\left(\frac{1}{\sqrt{n}}\right).$$

Hence we have

(5.4.20) $$V_0\left(\sum_{i=1}^{m} Y_i U_i\right) = p_2 + o\left(\frac{1}{\sqrt{n}}\right).$$

Since by (5.4.6)

$$E_0\left[Y_i^3\right] = -\frac{3c}{\tau_i} - \frac{c^3}{\tau_i^3} + \frac{3c^2 \operatorname{sgn} c}{2\tau_i^2 \sqrt{n}} + \frac{3c^4 \operatorname{sgn} c}{2\tau_i^4 \sqrt{n}} + o\left(\frac{1}{\sqrt{n}}\right),$$

it follows from (5.4.14) that

$$(5.4.21) \qquad E_0\left[Y_i^3 U_i^3\right] = -\frac{3c}{\tau_i^4} - \frac{c^3}{\tau_i^6} - \frac{3c^2 \operatorname{sgn} c}{2\tau_i^5 \sqrt{n}} + \frac{3c^4 \operatorname{sgn} c}{2\tau_i^7 \sqrt{n}} + o\left(\frac{1}{\sqrt{n}}\right).$$

Since by (5.4.15), (5.4.19) and (5.4.21) the third order cumulant of $Y_i U_i$ is given by $\mathcal{K}_3(Y_i U_i) = o(1/\sqrt{n})$, it follows that that of $\sum_{i=1}^m Y_i U_i$ is done by

$$(5.4.22) \qquad \mathcal{K}_{3,0}\left(\sum_{i=1}^m Y_i U_i\right) = o\left(\frac{1}{\sqrt{n}}\right).$$

Hence it follows from (5.4.16), (5.4.20) and (5.4.22) that the asymptotic expansion of the distribution of the estimator $\hat{\theta}_{WM}$ up to the order $n^{-1/2}$ is given by

$$(5.4.23) \quad P_{0,\tau_1,\ldots,\tau_m}^n\left\{\sqrt{n}\hat{\theta}_{WM} \le c\right\} = P_{0,\tau_1,\ldots,\tau_m}^n\left\{\sum_{i=1}^m Y_i U_i \le 0\right\}$$

$$= \Phi\left(c\sqrt{p_2}\right) - \frac{p_3 c^2}{2\sqrt{p_2 n}}\phi\left(c\sqrt{p_2}\right)\operatorname{sgn} c + o\left(\frac{1}{\sqrt{n}}\right).$$

We also have by (5.4.23)

$$(5.4.24) \quad P_{\theta,\tau_1,\ldots,\tau_m}^n\left\{\sqrt{p_2 n}\left(\hat{\theta}_{WM} - \theta\right) \le t\right\} = \Phi(t) - \frac{p_3 t^2}{2p_2^{3/2}\sqrt{n}}\phi(t)\operatorname{sgn} t + o\left(\frac{1}{\sqrt{n}}\right).$$

In a similar way to Akahira and Takeuchi (1981, page 97) it is shown that the bound for the asymptotic distribution of second order asymptotically median unbiased estimators based on the samples $\{X_{ij}\}$ is given by

$$\Phi(t) - \frac{p_3 t^2}{6p_2^{3/2}\sqrt{n}}\phi(t)\operatorname{sgn} t + o\left(\frac{1}{\sqrt{n}}\right) = F^*(t) \qquad \text{(say)},$$

that is, for any second order asymptotically median unbiased estimator $\hat{\theta}_n$

$$P_{\theta,\tau_1,\ldots,\tau_m}^n\left\{\sqrt{p_2 n}\left(\hat{\theta}_n - \theta\right) \le t\right\} \le F^*(t) \qquad \text{for all } t > 0;$$

$$P_{\theta,\tau_1,\ldots,\tau_m}^n\left\{\sqrt{p_2 n}\left(\hat{\theta}_n - \theta\right) \le t\right\} \ge F^*(t) \qquad \text{for all } t < 0.$$

Here an estimator $\hat{\theta}_n$ of θ based on the samples $\{X_{ij}\}$ is called second order asymptotically median unbiased (AMU) if

$$\lim_{n\to\infty} \sqrt{n}\left|P_{\theta,\tau_1,\ldots,\tau_m}^n\left\{\hat{\theta}_n \le \theta\right\} - \frac{1}{2}\right| = \lim_{n\to\infty} \sqrt{n}\left|P_{\theta,\tau_1,\ldots,\tau_m}^n\left\{\hat{\theta}_n \ge \theta\right\} - \frac{1}{2}\right| = 0$$

uniformly in some neighborhood of θ. From (5.4.9) and (5.4.24) it is easily seen that the estimators $\hat{\theta}_{ML}^0$ and $\hat{\theta}_{WM}$ are second order AMU. Thus we complete the proof.

Remark 5.4.2. In order to find the asymptotic distribution of the MLE $\hat{\theta}_{ML}^0$ up to the second order, i.e. the order $n^{-1/2}$, it is enough to show that the third order cumulant is of order n^{-1}, but the value given by (5.4.7) may be necessary in the case of higher order than second one.

Theorem 5.4.2. If τ_i $(i = 1,\ldots,m)$ are known, then the MLE $\hat{\theta}_{ML}^0$ is asymptotically better than $\hat{\theta}^0$ up to the order $n^{-1/2}$ in the sense that

$$\lim_{n\to\infty} \sqrt{n}\left[V_\theta\left(\sqrt{p_2 n}\left(\hat{\theta}^0 - \theta\right)\right) - V_\theta\left(\sqrt{p_2 n}\left(\hat{\theta}_{ML}^0 - \theta\right)\right)\right] \geq 0$$

Remark 5.4.3. By Remark 5.4.1 and Theorem 5.4.2 we see that the weighted median $\hat{\theta}_{WM}$ by the weights $1\big/\hat{\tau}_i\left(\hat{\theta}_{WM}\right)$ $(i = 1,\ldots,m)$ is also asymptotically better than $\hat{\theta}^0$ up to the order $n^{-1/2}$.

Proof. If $\tau_i = \tau$ $(i = 1,\ldots,m)$ and it is known, then the MLE $\hat{\theta}_{ML}^0$ is the median of $\{X_{ij}\}$. In a similar way to Akahira and Takeuchi (1981, page 97) it follows that the asymptotic distribution of $\hat{\theta}_{ML}^0$ up to the order $n^{-1/2}$ is given by

$$(5.4.25) \quad P_{\theta,\tau_1,\ldots,\tau_m}^n\left\{\frac{\sqrt{mn}}{\tau}\left(\hat{\theta}_{ML}^0 - \theta\right) \leq t\right\} = \Phi(t) - \frac{t^2}{2\sqrt{mn}}\phi(t)\mathrm{sgn}t + o\left(\frac{1}{\sqrt{n}}\right).$$

Hence it is seen that (5.4.25) is consistent with the asymptotic distribution of $\hat{\theta}_{ML}^0$ given in Theorem 5.4.1 in the same situation.

If τ_i $(i = 1,\ldots,m)$ are known, then for each $i = 1,\ldots,m$ the MLE $\hat{\theta}_i$ of θ is the median of X_{ij} $(j = 1,\ldots,n)$, i.e. $\hat{\theta}_i = \mathrm{med}_{1\leq j\leq n}X_{ij}$. Then it follows from (5.4.25) that for each $i = 1,\ldots,m$

$$(5.4.26) \quad P_{\theta,\tau_1,\ldots,\tau_m}^n\left\{\frac{\sqrt{n}}{\tau_i}\left(\hat{\theta}_i - \theta\right) \leq t\right\} = \Phi(t) - \frac{t^2}{2\sqrt{n}}\phi(t)\mathrm{sgn}t + o\left(\frac{1}{\sqrt{n}}\right)$$
$$= F_{\hat{\theta}_i}(t) \quad \text{(say)}.$$

From (5.4.26) we have the asymptotic density of $\hat{\theta}_i$ up to the order $n^{-1/2}$

$$(5.4.27) \quad f_{\hat{\theta}_i}(t) = \frac{d}{dt}F_{\hat{\theta}_i}(t) = \phi(t) + \frac{1}{2\sqrt{n}}\left(t^3 - 2t\right)\phi(t)\mathrm{sgn}t + o\left(\frac{1}{\sqrt{n}}\right).$$

From (5.4.27) we obtain as the asymptotic mean of $\hat{\theta}_i$

$$(5.4.28) \quad E_\theta\left[\frac{\sqrt{n}}{\tau_i}\left(\hat{\theta}_i - \theta\right)\right] = \int_{-\infty}^{\infty} t f_{\hat{\theta}_i}(t)dt = o\left(\frac{1}{\sqrt{n}}\right).$$

Hence it is seen by (5.4.28) that the asymptotic variance of $\hat{\theta}_i$ is given by

$$(5.4.29) \quad V_\theta\left(\frac{\sqrt{n}}{\tau_i}\left(\hat{\theta}_i - \theta\right)\right) = \int_{-\infty}^{\infty} t^2 f_{\hat{\theta}_i}(t)dt = 1 + \frac{2\sqrt{2}}{\sqrt{n\pi}} + o\left(\frac{1}{\sqrt{n}}\right).$$

Since

$$\hat{\theta}^0 = \sum_{i=1}^{m} \frac{1}{\tau_i^2} \hat{\theta}_i \bigg/ \sum_{i=1}^{m} \frac{1}{\tau_i^2} = \frac{1}{p_2} \sum_{i=1}^{m} \frac{1}{\tau_i^2} \hat{\theta}_i,$$

it follows from (5.4.29) that the asymptotic variance of $\hat{\theta}^0$ is given by

(5.4.30) $$V_\theta\left(\sqrt{p_2 n}\left(\hat{\theta}^0 - \theta\right)\right) = 1 + \frac{2\sqrt{2}}{\sqrt{n\pi}} + o\left(\frac{1}{\sqrt{n}}\right).$$

Since $\hat{\theta}_i$ $(i = 1, \ldots, m)$ are independent, it follows from (5.4.26) that the limiting distribution of $\sqrt{p_2 n}\left(\hat{\theta}^0 - \theta\right)$ is the standard normal distribution in the first order. On the other hand we have from (5.4.9)

(5.4.31) $$V_\theta\left(\sqrt{p_2 n}\left(\hat{\theta}^0_{ML} - \theta\right)\right) = 1 + \frac{p_3}{p_2^{3/2}} \cdot \frac{2\sqrt{2}}{\sqrt{n\pi}} + o\left(\frac{1}{\sqrt{n}}\right).$$

Note that $1/\sqrt{m} \le p_3/p_2^{3/2} \le 1$ since

$$\left\{\left(\sum_{i=1}^{m} \frac{1}{\tau_i^2}\right)\bigg/ m\right\}^{1/2} \le \left\{\left(\sum_{i=1}^{m} \frac{1}{\tau_i^3}\right)\bigg/ m\right\}^{1/3}; \qquad \left(\sum_{i=1}^{m} \frac{1}{\tau_i^3}\right)^2 \le \left(\sum_{i=1}^{m} \frac{1}{\tau_i^2}\right)^3.$$

Then we obtain by (5.4.30) and (5.4.31)

(5.4.32) $$\lim_{n \to \infty} \sqrt{n}\left[V_\theta\left(\sqrt{p_2 n}\left(\hat{\theta}^0 - \theta\right)\right) - V_\theta\left(\sqrt{p_2 n}\left(\hat{\theta}^0_{ML} - \theta\right)\right)\right]$$
$$= \frac{2\sqrt{2}}{\sqrt{\pi}}\left(1 - \frac{p_3}{p_2^{3/2}}\right) \ge \frac{2\sqrt{2}}{\sqrt{\pi}}\left(1 - \frac{1}{\sqrt{m}}\right) \ge 0.$$

Thus we complete the proof.

Theorem 5.4.3. *The weighted mean* $\hat{\theta}^*$ *of* $\hat{\theta}_i$ $(i = 1, \ldots, m)$ *by the weights* $1\big/\hat{\tau}_i^2\left(\hat{\theta}_i\right)$ $(i = 1, \ldots, m)$ *is asymptotically equivalent to the weighted mean* $\hat{\theta}^0$ *by the weights* $1/\tau_i^2$ $(i = 1, \ldots, m)$ *up to the order* $n^{-1/2}$ *in the sense that*

$$\lim_{n \to \infty} \sqrt{n}\left|V_\theta\left(\sqrt{p_2 n}\left(\hat{\theta}^* - \theta\right)\right) - V_\theta\left(\sqrt{p_2 n}\left(\hat{\theta}^0 - \theta\right)\right)\right| = 0.$$

Proof. We shall obtain the asymptotic variance of the weighted mean $\hat{\theta}^*$ of $\hat{\theta}_i$ $(i = 1, \ldots, m)$ by the weights $1\big/\hat{\tau}_i^2\left(\hat{\theta}_i\right)$ $(i = 1, \ldots, m)$, where $\hat{\tau}_i\left(\hat{\theta}_i\right) = \sum_{j=1}^{n}\left|X_{ij} - \hat{\theta}_i\right|$ with $\hat{\theta}_i = \text{med}_{1 \le j \le n} X_{ij}$ $(i = 1, \ldots, m)$. Putting $\Delta_i = \hat{\tau}_i - \tau_i$ $(i = 1, \ldots, m)$, we have

(5.4.33) $$\sum_{i=1}^{m} \frac{1}{\hat{\tau}_i^2} = \sum_{i=1}^{m} \frac{1}{\tau_i^2\left\{1 + (1/\tau_i)(\hat{\tau}_i - \tau_i)\right\}^2} = \sum_{i=1}^{m} \frac{1}{\tau_i^2}\left(1 - \frac{2\Delta_i}{\tau_i} + o_p(\Delta_i)\right).$$

By (5.4.33) we obtain

$$
(5.4.34) \quad \hat{\theta}^* = \sum_{i=1}^m \frac{1}{\hat{\tau}_i^2} \hat{\theta}_i \bigg/ \sum_{i=1}^m \frac{1}{\hat{\tau}_i^2}
$$

$$
= \frac{1}{p_2} \left(\sum_{i=1}^m \frac{\hat{\theta}_i}{\tau_i^2} - 2 \sum_{i=1}^m \frac{\hat{\theta}_i \Delta_i}{\tau_i^3} + \frac{2}{p_2} \sum_{i=1}^m \frac{\hat{\theta}_i}{\tau_i^2} \sum_{i=1}^m \frac{\Delta_i}{\tau_i^3} \right) + o_p \left(\sum_{i=1}^m \Delta_i \right).
$$

Without loss of generality we assume that $\theta = 0$. Since by (5.4.34)

$$
(5.4.35) \qquad \sqrt{n}\hat{\theta}^* = \frac{1}{p_2} \left(\sum_{i=1}^m \frac{\sqrt{n}\hat{\theta}_i}{\tau_i^2} - \frac{2}{\sqrt{n}} \sum_{i=1}^m \frac{\sqrt{n}\hat{\theta}_i \sqrt{n}\Delta_i}{\tau_i^3} \right.
$$

$$
\left. + \frac{2}{p_2\sqrt{n}} \sum_{i=1}^m \frac{\sqrt{n}\hat{\theta}_i}{\tau_i^2} \sum_{i=1}^m \frac{\sqrt{n}\Delta_i}{\tau_i^3} \right) + o_p \left(\frac{1}{\sqrt{n}} \right),
$$

it follows from (5.4.28) that

$$
(5.4.36) \quad E_0 \left[\sqrt{n}\hat{\theta}^* \right] = \frac{1}{p_2} \left\{ \sum_{i=1}^m \frac{1}{\tau_i^2} E_0 \left[\sqrt{n}\hat{\theta}_i \right] - \frac{2}{\sqrt{n}} \sum_{i=1}^m \frac{1}{\tau_i^3} E_0 \left[\sqrt{n}\hat{\theta}_i E_0 \left[\sqrt{n}\Delta_i \big| \hat{\theta}_i \right] \right] \right.
$$

$$
\left. + \frac{2}{p_2\sqrt{n}} \sum_{i=1}^m \sum_{j=1}^m \frac{1}{\tau_i^2 \tau_j^3} E_0 \left[\sqrt{n}\hat{\theta}_i E_0 \left[\sqrt{n}\Delta_i \big| \hat{\theta}_i \right] \right] \right\}
$$

$$
+ o \left(\frac{1}{\sqrt{n}} \right),
$$

where $E[\cdot|\hat{\theta}_i]$ denotes the asymptotic conditional mean given $\hat{\theta}_i$. After some manipulation we have by (5.4.36)

$$
(5.4.37) \qquad E_0 \left[\sqrt{n}\hat{\theta}^* \right] = o \left(\frac{1}{\sqrt{n}} \right).
$$

Since by (5.4.35)

$$
E_0 \left[n\hat{\theta}^{*2} \right] = \frac{1}{p_2^2} \left\{ E_0 \left[\left(\sum_{i=1}^m \frac{\sqrt{n}\hat{\theta}_i}{\tau_i^2} \right)^2 \right] - \frac{4}{\sqrt{n}} E_0 \left[\sum_{i=1}^m \frac{\sqrt{n}\hat{\theta}_i}{\tau_i} E_0 \left[\sum_{i=1}^m \frac{\sqrt{n}\hat{\theta}_i \sqrt{n}\Delta_i}{\tau_i^3} \bigg| \hat{\theta}_i \right] \right] \right.
$$

$$
\left. + \frac{4}{p_2\sqrt{n}} E_0 \left[\left(\sum_{i=1}^m \frac{\sqrt{n}\hat{\theta}_i}{\tau_i^2} \right)^2 E_0 \left[\sum_{i=1}^m \frac{\sqrt{n}\Delta_i}{\tau_i} \bigg| \hat{\theta}_i \right] \right] \right\} + o_p \left(\frac{1}{\sqrt{n}} \right),
$$

it follows that

$$
(5.4.38) \qquad E_0 \left[n\hat{\theta}^{*2} \right] = \frac{1}{p_2} \left\{ 1 + \frac{2\sqrt{2}}{\sqrt{n\pi}} \right\} + O \left(\frac{1}{n} \right).
$$

From (5.4.37) and (5.4.38) we have

$$(5.4.39) \qquad V_0 \left(\sqrt{p_2 n} \hat{\theta}^* \right) = 1 + \frac{2\sqrt{2}}{\sqrt{n\pi}} + O \left(\frac{1}{n} \right).$$

From (5.4.30) and (5.4.39) the conclusion follows as required.

Remark 5.4.4. It is noted from (5.4.26) and (5.4.35) that the limiting distribution of $\sqrt{p_2 n} \hat{\theta}^*$ is the standard normal distribution in the first order.

Corollary 5.4.1. *The weighted median $\hat{\theta}_{WM}$ by the weights $1 \big/ \hat{r}_i \left(\hat{\theta}_{WM} \right)$ ($i = 1, \ldots, m$) is asymptotically better than the weighted mean $\hat{\theta}^*$ up to the order $n^{-1/2}$ in the sense that*

$$\lim_{n \to \infty} \sqrt{n} \left[V_\theta \left(\sqrt{p_2 n} \left(\hat{\theta}_{WM} - \theta \right) \right) - V_\theta \left(\sqrt{p_2 n} \left(\hat{\theta}^* - \theta \right) \right) \right] \geq 0.$$

The proof is straightforward from Remark 5.4.3 and Theorem 5.4.3.

As is seen from the above discussion, the second order asymptotic comparison of the asymptotically efficient estimators $\hat{\theta}_{ML}^0$, $\hat{\theta}_{WM}$, $\hat{\theta}^0$ and $\hat{\theta}^*$ is given by

$$\hat{\theta}_{WM} \sim \hat{\theta}_{ML}^0 \succ \hat{\theta}^0 \sim \hat{\theta}^*,$$

where "$a \sim b$" ("$a \succ b$") means that a is asymptotically equivalent to (better than) b up to the order $n^{-1/2}$.

CHAPTER 6

HIGHER ORDER ASYMPTOTICS IN ESTIMATION FOR TWO-SIDED WEIBULL TYPE DISTRIBUTIONS

Higher order asymptotics has been studied by Pfanzagl and Wefelmeyer (1985), Ghosh et al. (1980), Akahira and Takeuchi (1981), Akahira (1986), Akahira et al. (1988), Ghosh (1994) among others, under suitable regularity conditions.

In non-regular cases when the regularity conditions do not necessarily hold, the (higher order) asymptotics was discussed by Daniels (1961), Williamson (1984), Ibragimov and Has'minskii (1981), Akahira and Takeuchi (1981), Akahira (1987, 1988a, 1988b), Pfanzagl and Wefelmeyer (1985) and others.

In this chapter we consider the estimation problem of a location parameter θ on a sample of size n from a two-sided Weibull type density $f(x-\theta) = C(\alpha) \exp\left(-|x-\theta|^\alpha\right)$ for $-\infty < x < \infty$, $-\infty < \theta < \infty$ and $1 < \alpha < 3/2$, where $C(\alpha) = \alpha/\{2\Gamma(1/\alpha)\}$. It is noted that there is a Fisher information amount and a first order derivative of $f(x)$ at $x = 0$, but there is no second order one of $f(x)$ at $x = 0$. It is also seen in Akahira (1975b) that the order of consistency is equal to $n^{1/2}$ in this situation. Then we shall obtain the bound for the distribution of asymptotically median unbiased estimators of θ up to the 2α-th order, i.e., the order $n^{-(2\alpha-1)/2}$. We shall also get the asymptotic distribution of the maximum likelihood estimator (MLE) of θ up to the 2α-th order and see that the MLE is not generally 2α-th order asymptotically efficient. Further, we shall obtain the amount of the loss of asymptotic information of the MLE.

6.1. THE 2α-TH ORDER ASYMPTOTIC BOUND FOR THE DISTRIBUTION OF 2α-TH ORDER AMU ESTIMATORS

Let $X_1, \ldots, X_n, \ldots$ be a sequence of independent and identically distributed (i.i.d.) random variables with a two-sided Weibull type density $f(x-\theta) = C(\alpha) \exp\left\{-|x-\theta|^\alpha\right\}$ for $-\infty < x < \infty$ where θ is a real-valued parameter, $1 < \alpha < 3/2$ and $C(\alpha) = \alpha/\{2\Gamma(1/\alpha)\}$ with a Gamma function $\Gamma(u)$, i.e., $\Gamma(u) = \int_0^\infty x^{u-1} e^{-x} dx$ $(u > 0)$.

We denote by P_θ^n the n-fold direct products of probability measure P_θ with the above density $f(x - \theta)$. An estimator $\hat{\theta}_n$ of θ based on $X_1, \ldots, X_n$ is called a 2α-th order asymptotically median unbiased (AMU) estimator if for any $\eta \in \mathbf{R}^1$, there exists a positive number δ such that

$$\lim_{n \to \infty} \sup_{\theta \,:\, |\theta - \eta| < \delta} n^{(2\alpha - 1)/2} \left| P_\theta^n \left\{ \hat{\theta}_n \leq \theta \right\} - \frac{1}{2} \right| = 0,$$

$$\lim_{n \to \infty} \sup_{\theta \,:\, |\theta - \eta| < \delta} n^{(2\alpha - 1)/2} \left| P_\theta^n \left\{ \hat{\theta}_n \geq \theta \right\} - \frac{1}{2} \right| = 0.$$

We denote by $A_{2\alpha}$ the class of all best asymptotically normal and 2α-th order AMU estimators. For a $\hat{\theta}_n$ 2α-th order AMU, $G_0(t, \theta) + n^{-(2\alpha - 1)/2} G_1(t, \theta)$ is defined to be the 2α-th order asymptotic distribution of $\sqrt{n}\left(\hat{\theta}_n - \theta\right)$ (or $\hat{\theta}_n$ for short) if there exist a continuous function $G_0(\cdot, \theta)$ and an absolutely continuous function $G_1(\cdot, \theta)$ such that for any $t \in \mathbf{R}^1$ and each $\theta \in \mathbf{R}^1$

$$\lim_{n \to \infty} n^{(2\alpha - 1)/2} \left| P_\theta^n \left\{ \sqrt{n}\left(\hat{\theta}_n - \theta\right) \leq t \right\} - G_0(t, \theta) - n^{-(2\alpha - 1)/2} G_1(t, \theta) \right| = 0.$$

In order to obtain the bound for the distribution of 2α-th order AMU estimators of θ, for arbitrary but fixed θ_0, we consider the problem of testing hypothesis $H : \theta = \theta_0 + tn^{-1/2}$ $(t > 0)$ against the alternative $K : \theta = \theta_0$. Then the log-likelihood ratio test statistic Z_n is given by

$$Z_n = \sum_{i=1}^{n} \log \left\{ f(X_i - \theta_0) \Big/ f\left(X_i - \theta_0 - tn^{-1/2}\right) \right\}$$

$$= -\sum_{i=1}^{n} \left(|X_i - \theta_0|^\alpha - \left| X_i - \theta_0 - tn^{-1/2} \right|^\alpha \right).$$

In order to obtain the asymptotic cumulants of Z_n, we need the following lemma.

Lemma 6.1.1. *If* $h_\Delta(x) = (x + \Delta)^\alpha - x^\alpha$ *for* $\Delta > 0$, *then*

$$\int_0^\infty h_\Delta^2(x) e^{-x^\alpha} dx = \alpha \Gamma\left(2 - \frac{1}{\alpha}\right) \Delta^2 + \alpha(\alpha - 1) \Gamma\left(2 - \frac{2}{\alpha}\right) \Delta^3$$
$$- \frac{1 + \gamma}{2\alpha + 1} \Delta^{2\alpha + 1} + o\left(\Delta^{2\alpha + 1}\right),$$

$$\int_0^\infty h_\Delta^3(x) e^{-x^\alpha} dx = \alpha^2 \Gamma\left(3 - \frac{2}{\alpha}\right) \Delta^3 + O\left(\Delta^4\right),$$

where

$$\gamma = \frac{\alpha(\alpha - 1)\Gamma(\alpha - 1)\Gamma(3 - 2\alpha)}{2(2\alpha - 1)\Gamma(2 - \alpha)}.$$

The proof is given in Section 6.2.

In the following lemma we obtain the asymptotic mean, variance and third-order cumulant of Z_n, under H and K.

Lemma 6.1.2. *The asymptotic mean, variance and third-order cumulant of Z_n are given as follows : Under $K : \theta = \theta_0$,*

$$E_{\theta_0}(Z_n) = \frac{I}{2}t^2 - \frac{k}{2}t^{2\alpha+1}n^{-(2\alpha-1)/2} + o\left(n^{-(2\alpha-1)/2}\right),$$

$$V_{\theta_0}(Z_n) = It^2 - kt^{2\alpha+1}n^{-(2\alpha-1)/2} + o\left(n^{-(2\alpha-1)/2}\right),$$

$$\mathcal{K}_{3,\theta_0}(Z_n) = o\left(n^{-(2\alpha-1)/2}\right),$$

and under $H : \theta = \theta_0 + tn^{-1/2}$,

$$E_{\theta_0+tn^{-1/2}}(Z_n) = -\frac{I}{2}t^2 + \frac{k}{2}t^{2\alpha+1}n^{-(2\alpha-1)/2} + o\left(n^{-(2\alpha-1)/2}\right),$$

$$V_{\theta_0+tn^{-1/2}}(Z_n) = It^2 - kt^{2\alpha+1}n^{-(2\alpha-1)/2} + o\left(n^{-(2\alpha-1)/2}\right),$$

$$\mathcal{K}_{3,\theta_0+tn^{-1/2}}(Z_n) = o\left(n^{-(2\alpha-1)/2}\right),$$

where

$$I = E_\theta\left[\left\{\frac{\partial}{\partial\theta}\log f(X-\theta)\right\}^2\right]$$

$$= -E_\theta\left[\frac{\partial^2}{\partial\theta^2}\log f(X-\theta)\right] = \alpha(\alpha-1)\Gamma(1-(1/\alpha))/\Gamma(1/\alpha)$$

and

$$k = \alpha\left\{B(\alpha+1,\alpha+1) + \frac{\gamma}{2\alpha+1}\right\}\Big/\Gamma(1/\alpha)$$

with

$$B(u,v) = \int_0^1 x^{u-1}(1-x)^{v-1}dx \qquad (u,v>0).$$

The proof is given in Section 6.2.

In order to get the bound for the 2α-th order asymptotic distribution of 2α-th order AMU estimators, we need the following.

Lemma 6.1.3. *Assume that the asymptotic mean, variance and third-order cumulant of Z_n, under the distributions $P_{\theta,n}$ for $\theta = \theta_0 + tn^{-1/2}$, are given by the following form.*

$$E_\theta(Z_n) \doteq \mu(t,\theta) + n^{-(2\alpha-1)/2}C_1(t,\theta) + o\left(n^{-(2\alpha-1)/2}\right),$$

$$V_\theta(Z_n) \doteq v^2(t,\theta) + n^{-(2\alpha-1)/2}C_2(t,\theta) + o\left(n^{-(2\alpha-1)/2}\right),$$

$$\mathcal{K}_{3,\theta}(Z_n) = o\left(n^{-(2\alpha-1)/2}\right).$$

Then

$$P_\theta^n\{Z_n \le \alpha_0\} = \frac{1}{2} + o\left(n^{-(2\alpha-1)/2}\right)$$

if and only if

$$\alpha_0 = \mu(t, \theta) + C_1(t, \theta) n^{-(2\alpha-1)/2} + o\left(n^{-(2\alpha-1)/2}\right).$$

The proof is essentially given in Akahira and Takeuchi ((1981), pp. 132, 133). In the following theorem we obtain the 2α-th asymptotic bound for the distribution of 2α-th order AMU estimators of θ.

Theorem 6.1.1. *The bound for the 2α-th order asymptotic distribution of 2α-th order AMU estimators of θ is given by*

$$\Phi(t) - C_0|t|^{2\alpha}\phi(t) n^{-(2\alpha-1)/2}\mathrm{sgn}t + o\left(n^{-(2\alpha-1)/2}\right),$$

that is, for any $\hat{\theta}_n \in A_{2\alpha}$

$$P_\theta^n\left\{\sqrt{In}\left(\hat{\theta}_n - \theta\right) \le t\right\} \le \Phi(t) - C_0 t^{2\alpha}\phi(t) n^{-(2\alpha-1)/2} + o\left(n^{-(2\alpha-1)/2}\right)$$

$$\text{for all} \quad t > 0,$$

$$P_\theta^n\left\{\sqrt{In}\left(\hat{\theta}_n - \theta\right) \le t\right\} \ge \Phi(t) + C_0|t|^{2\alpha}\phi(t) n^{-(2\alpha-1)/2} + o\left(n^{-(2\alpha-1)/2}\right)$$

$$\text{for all} \quad t < 0,$$

where

$$C_0 = \frac{\alpha\left\{B(\alpha+1, \alpha+1) + (\gamma/(2\alpha+1))\right\}}{2I^{\alpha+(1/2)}\Gamma(1/\alpha)}$$

and $\Phi(t)$ and $\phi(t)$ denote the standard normal distribution function and its density function, respectively.

The proof is given in Section 6.2.

Remark 6.1.1. The result of Theorem 6.1.1 holds for $2/3 < \alpha < 1$, where the information amount I must be expressed as $\alpha^2\Gamma(2 - (1/\alpha))/\Gamma(1/\alpha)$. The proof is omitted since it is essentially similar to the above.

6.2. PROOFS OF LEMMAS AND THEOREM IN SECTION 6.1

In this section we gives the proofs of Lemmas 6.1.1, 6.1.2, 6.1.3 and Theorem 6.1.1.
Proof of Lemma 6.1.1. First we have

$$(6.2.1) \quad \int_0^\infty h_\Delta^2(x)e^{-x^\alpha}\,dx = \int_0^\infty (x+\Delta)^{2\alpha}e^{-x^\alpha}\,dx - 2\int_0^\infty (x+\Delta)^\alpha x^\alpha e^{-x^\alpha}\,dx$$

$$+ \int_0^\infty x^{2\alpha}e^{-x^\alpha}\,dx.$$

Since for $\beta > 0$

$$(6.2.2) \quad \int_0^\infty x^{\beta-1}e^{-x^\alpha}\,dx = \frac{1}{\alpha}\Gamma\left(\frac{\beta}{\alpha}\right),$$

it follows that

$$(6.2.3) \int_0^\infty (x+\Delta)^{2\alpha} e^{-x^\alpha} dx = -\frac{\Delta^{2\alpha+1}}{2\alpha+1} + \frac{\alpha}{2\alpha+1} \int_0^\infty (x+\Delta)^{2\alpha+1} x^{\alpha-1} e^{-x^\alpha} dx$$

$$= -\frac{\Delta^{2\alpha+1}}{2\alpha+1} + \frac{\alpha}{2\alpha+1} \int_0^\infty \Big\{ x^{3\alpha} + (2\alpha+1)x^{3\alpha-1}\Delta$$

$$+ \alpha(2\alpha+1)x^{3\alpha-2}\Delta^2 + \frac{1}{3}\alpha(2\alpha-1)(2\alpha+1)x^{3\alpha-3}\Delta^3 \Big\}$$

$$\cdot e^{-x^\alpha} dx + O\left(\Delta^4\right)$$

$$= \frac{1}{2\alpha+1}\Gamma\left(3+\frac{1}{\alpha}\right) + 2\Delta + \alpha\Gamma\left(3-\frac{1}{\alpha}\right)\Delta^2$$

$$+ \frac{1}{3}\alpha(2\alpha-1)\Gamma\left(3-\frac{2}{\alpha}\right)\Delta^3 - \frac{1}{2\alpha+1}\Delta^{2\alpha+1} + O\left(\Delta^4\right).$$

From (6.2.2), we obtain

$$\int_0^\infty (x+\Delta)^\alpha x^\alpha e^{-x^\alpha} dx = -\frac{\alpha}{\alpha+1} \int_0^\infty (x+\Delta)^{\alpha+1} \left(x^{\alpha-1} - x^{2\alpha-1}\right) e^{-x^\alpha} dx$$

$$= \frac{\alpha}{\alpha+1} \int_0^\infty \Big\{ x^{\alpha+1} + (\alpha+1)x^\alpha\Delta + \frac{1}{2}\alpha(\alpha+1)x^{\alpha-1}\Delta^2$$

$$+ \frac{1}{6}\alpha(\alpha-1)(\alpha+1)x^{\alpha-2}\Delta^3 \Big\} \left(x^{2\alpha-1} - x^{\alpha-1}\right) e^{-x^\alpha} dx$$

$$- \frac{\alpha}{\alpha+1} \int_0^\infty R(\Delta)x^{\alpha-1} e^{-x^\alpha} dx + O\left(\Delta^4\right),$$

where

$$R(\Delta) = (x+\Delta)^{\alpha+1} - x^{\alpha+1} - (\alpha+1)\Delta x^\alpha - \frac{1}{2}\alpha(\alpha+1)\Delta^2 x^{\alpha-1}$$

$$- \frac{1}{6}\alpha(\alpha-1)(\alpha+1)\Delta^3 x^{\alpha-2}.$$

Then the remainder term $R(\Delta)$ of the Taylor expansion is represented by

$$R(\Delta) = K_\alpha \int_0^\Delta (\Delta-t)^3 (x+t)^{\alpha-3} dt,$$

where $0 \leq t \leq \Delta$ and $K_\alpha = \alpha(\alpha+1)(\alpha-1)(\alpha-2)/6$. Since $1 - x^\alpha < e^{-x^\alpha} < 1$, it follows that

$$\int_0^\infty (x+t)^{\alpha-3} x^{\alpha-1} \left(1 - e^{-x^\alpha}\right) dx \leq \int_0^\infty x^{2\alpha-4} \left(1 - e^{-x^\alpha}\right) dx$$

$$= \int_0^1 x^{2\alpha-4} \left(1 - e^{-x^\alpha}\right) dx$$

$$+ \int_1^\infty x^{2\alpha-4} \left(1 - e^{-x^\alpha}\right) dx$$

$$\leq \int_0^1 x^{3\alpha-4}dx + \int_1^\infty x^{2\alpha-4}dx$$

$$= \frac{\alpha}{3(\alpha-1)(3-2\alpha)}.$$

Since $\int_0^\infty (x+t)^{\alpha-3}x^{\alpha-1}dx = t^{2\alpha-3}B(\alpha,3-2\alpha)$, we have

$$\int_0^\infty R(\Delta)x^{\alpha-1}e^{-z^\alpha}dx = \int_0^\infty R(\Delta)x^{\alpha-1}dx - \int_0^\infty R(\Delta)x^{\alpha-1}\left(1-e^{-z^\alpha}\right)dx$$

$$= K_\alpha \int_0^\Delta (\Delta-t)^3\left\{\int_0^\infty (x+t)^{\alpha-3}x^{\alpha-1}dx\right\}dt + o\left(\Delta^{2\alpha+1}\right)$$

$$= K_\alpha B(\alpha,3-2\alpha)\int_0^\Delta t^{2\alpha-3}(\Delta-t)^3 dt + o\left(\Delta^{2\alpha+1}\right)$$

$$= K_\alpha B(\alpha,3-2\alpha)B(2\alpha-2,4)\Delta^{2\alpha+1} + o\left(\Delta^{2\alpha+1}\right)$$

$$= -\frac{(\alpha+1)(\alpha-1)\Gamma(\alpha-1)\Gamma(3-2\alpha)}{4(2\alpha+1)(2\alpha-1)\Gamma(2-\alpha)}\Delta^{2\alpha+1} + o\left(\Delta^{2\alpha+1}\right).$$

Hence, we obtain

$$(6.2.4)\quad \int_0^\infty (x+\Delta)^\alpha x^\alpha e^{-z^\alpha}dx = \frac{1}{\alpha+1}\left\{\Gamma\left(3+\frac{1}{\alpha}\right)-\Gamma\left(2+\frac{1}{\alpha}\right)\right\}$$

$$+\Delta+\frac{\alpha}{2}\left\{\Gamma\left(3-\frac{1}{\alpha}\right)-\Gamma\left(2-\frac{1}{\alpha}\right)\right\}\Delta^2$$

$$+\frac{1}{6}\alpha(\alpha-1)\left\{\Gamma\left(3-\frac{2}{\alpha}\right)-\Gamma\left(2-\frac{2}{\alpha}\right)\right\}\Delta^3$$

$$+\frac{\alpha(\alpha-1)\Gamma(\alpha-1)\Gamma(3-2\alpha)}{4(2\alpha+1)(2\alpha-1)\Gamma(2-\alpha)}\Delta^{2\alpha+1} + o\left(\Delta^{2\alpha+1}\right)$$

$$= \frac{1}{\alpha}\Gamma\left(2+\frac{1}{\alpha}\right)+\Delta+\left(\frac{\alpha-1}{2}\right)\Gamma\left(2-\frac{1}{\alpha}\right)\Delta^2$$

$$+\frac{1}{6}(\alpha-1)(\alpha-2)\Gamma\left(2-\frac{2}{\alpha}\right)\Delta^3$$

$$+\frac{\alpha(\alpha-1)\Gamma(\alpha-1)\Gamma(3-2\alpha)}{4(2\alpha+1)(2\alpha-1)\Gamma(2-\alpha)}\Delta^{2\alpha+1} + o\left(\Delta^{2\alpha+1}\right),$$

and, by (6.2.2), $\int_0^\infty x^{2\alpha}e^{-z^\alpha}dx = \Gamma(2+1/\alpha)/\alpha$. From (6.2.1) to (6.2.4), we have

$$\int_0^\infty h_\Delta^2(x)e^{-z^\alpha}dx = \alpha\Gamma\left(2-\frac{1}{\alpha}\right)\Delta^2 + \alpha(\alpha-1)\Gamma\left(2-\frac{2}{\alpha}\right)\Delta^3$$

$$-\frac{1+\gamma}{2\alpha+1}\Delta^{2\alpha+1} + o\left(\Delta^{2\alpha+1}\right),$$

where $\gamma = \alpha(\alpha - 1)\Gamma(\alpha - 1)\Gamma(3 - 2\alpha)/\{2(2\alpha - 1)\Gamma(2 - \alpha)\}$. We also obtain

$$\int_0^\infty h_\Delta^3(x)e^{-x^\alpha}\,dx = \int_0^\infty \{(x + \Delta)^\alpha - x^\alpha\}^3\,e^{-x^\alpha}\,dx$$

$$= \alpha^3\Delta^3 \int_0^\infty x^{3\alpha-3}e^{-x^\alpha}\,dx + O\left(\Delta^4\right)$$

$$= \alpha^2\Gamma\left(3 - \frac{2}{\alpha}\right)\Delta^3 + O\left(\Delta^4\right).$$

Thus we complete the proof.

Proof of Lemma 6.1.2. Without loss of generality, we assume that $\theta_0 = 0$. Putting $\Psi_\Delta(x) = |x|^\alpha - |x - \Delta|^\alpha$ with $\Delta > 0$, we have $Z_n = -\sum_{i=1}^n \Psi_\Delta(X_i)$ where $\Delta = tn^{-1/2}$. Since

(6.2.5) $$\Psi_\Delta(x) = \begin{cases} x^\alpha - (x - \Delta)^\alpha & \text{for} \quad x \geq \Delta, \\ x^\alpha - (\Delta - x)^\alpha & \text{for} \quad 0 \leq x < \Delta, \\ (-x)^\alpha - (\Delta - x)^\alpha & \text{for} \quad x < 0, \end{cases}$$

it follows that

(6.2.6) $$E_0\left[\Psi_\Delta(X)\right] = C(\alpha)\left[-\int_0^\infty \{(x + \Delta)^\alpha - x^\alpha\}\,e^{-x^\alpha}\,dx \right.$$

$$+ \int_0^\Delta \{x^\alpha - (\Delta - x)^\alpha\}\,e^{-x^\alpha}\,dx$$

$$\left. + \int_\Delta^\infty \{x^\alpha - (x - \Delta)^\alpha\}\,e^{-x^\alpha}\,dx \right]$$

$$= C(\alpha)\left(I_1 + I_2 + I_3\right) \qquad \text{(say)}.$$

Putting $h_\Delta(x) = (x + \Delta)^\alpha - x^\alpha$, we have from Lemma 6.1.1

(6.2.7) $$I_1 + I_3 = -\int_0^\infty \{(x + \Delta)^\alpha - x^\alpha\}\,e^{-x^\alpha}\,dx + \int_\Delta^\infty \{x^\alpha - (x - \Delta)^\alpha\}\,e^{-x^\alpha}\,dx$$

$$= \int_0^\infty \{(x + \Delta)^\alpha - x^\alpha\}\left\{e^{-(x+\Delta)^\alpha} - e^{-x^\alpha}\right\}\,dx$$

$$= \int_0^\infty h_\Delta(x)\left\{e^{-h_\Delta(x)} - 1\right\}e^{-x^\alpha}\,dx$$

$$= -\int_0^\infty h_\Delta^2(x)e^{-x^\alpha}\,dx + \frac{1}{2}\int_0^\infty h_\Delta^3(x)e^{-x^\alpha}\,dx + O\left(\Delta^4\right)$$

$$= -\alpha\Gamma\left(2 - \frac{1}{\alpha}\right)\Delta^2 + \frac{1 + \gamma}{2\alpha + 1}\Delta^{2\alpha+1} + o\left(\Delta^{2\alpha+1}\right).$$

We also obtain

$$(6.2.8) \qquad I_2 = \int_0^\Delta \left\{ x^\alpha - (\Delta - x)^\alpha \right\} e^{-x^\alpha} dx$$

$$= \int_0^\Delta \left\{ x^\alpha - (\Delta - x)^\alpha \right\} \left\{ 1 - x^\alpha + \frac{1}{2} x^{2\alpha} + O\left(x^{3\alpha} \right) \right\} dx$$

$$= -\frac{\Delta^{2\alpha+1}}{2\alpha+1} + \Delta^{2\alpha+1} \int_0^\Delta \left(\frac{x}{\Delta} \right)^\alpha \left(1 - \frac{x}{\Delta} \right)^\alpha \frac{1}{\Delta} dx + O\left(\Delta^{3\alpha+1} \right)$$

$$= \left\{ B(\alpha+1, \alpha+1) - \frac{1}{2\alpha+1} \right\} \Delta^{2\alpha+1} + O\left(\Delta^{3\alpha+1} \right),$$

where $B(u, v)$ denotes the Beta function. From (6.2.7) and (6.2.8), we have

$$I_1 + I_2 + I_3 = -\alpha\Gamma\left(2 - \frac{1}{\alpha} \right) \Delta^2$$

$$+ \left\{ B(\alpha+1, \alpha+1) + \frac{\gamma}{2\alpha+1} \right\} \Delta^{2\alpha+1} + o\left(\Delta^{2\alpha+1} \right).$$

Since $C(\alpha) = \alpha/\{2\Gamma(1/\alpha)\}$, it follows from (6.2.6), (6.2.7) and (6.2.8) that

$$(6.2.9) \quad E_0 \left[\Psi_\Delta(X) \right] = C(\alpha)\left(I_1 + I_2 + I_3 \right)$$

$$= -\frac{\alpha(\alpha-1)\Gamma\left(1 - \frac{1}{\alpha} \right)}{2\Gamma\left(\frac{1}{\alpha} \right)} \Delta^2$$

$$+ \frac{\alpha\left\{ B(\alpha+1, \alpha+1) + (\gamma/(2\alpha+1)) \right\}}{2\Gamma\left(\frac{1}{\alpha} \right)} \Delta^{2\alpha+1} + o\left(\Delta^{2\alpha+1} \right).$$

From (6.2.5), we obtain

$$(6.2.10) \quad E_0 \left[\Psi_\Delta^2(X) \right] = \int_{-\infty}^0 \left\{ (-x)^\alpha - (\Delta - x)^\alpha \right\}^2 f(x) dx + \int_0^\Delta \left\{ x^\alpha - (\Delta - x)^\alpha \right\}^2 f(x) dx$$

$$+ \int_\Delta^\infty \left\{ x^\alpha - (x - \Delta)^\alpha \right\}^2 f(x) dx$$

$$= \int_0^\infty \left\{ x^\alpha - (x + \Delta)^\alpha \right\}^2 f(x) dx + \int_0^\Delta \left\{ x^\alpha - (\Delta - x)^\alpha \right\}^2 f(x) dx$$

$$+ \int_\Delta^\infty \left\{ x^\alpha - (x - \Delta)^\alpha \right\}^2 f(x) dx$$

$$= C(\alpha)\left(I_1' + I_2' + I_3' \right) \qquad \text{(say)}.$$

Since $h_\Delta(x) = (x + \Delta)^\alpha - x^\alpha$, it follows from Lemma 6.1.1 that

$$(6.2.11) \quad I_1' + I_3' = \int_0^\infty \left\{x^\alpha - (x + \Delta)^\alpha\right\}^2 e^{-x^\alpha}\, dx + \int_\Delta^\infty \left\{x^\alpha - (x - \Delta)^\alpha\right\}^2 e^{-x^\alpha}\, dx$$

$$= \int_0^\infty \left\{(x + \Delta)^\alpha - x^\alpha\right\}^2 \left\{e^{-x^\alpha} + e^{-(x+\Delta)^\alpha}\right\} dx$$

$$= \int_0^\infty h_\Delta^2(x) \left\{e^{-h_\Delta(x)} + 1\right\} e^{-x^\alpha}\, dx$$

$$= 2\int_0^\infty h_\Delta^2(x) e^{-x^\alpha}\, dx - \int_0^\infty h_\Delta^3(x) e^{-x^\alpha}\, dx + O\left(\Delta^4\right)$$

$$= 2\alpha\Gamma\left(2 - \frac{1}{\alpha}\right)\Delta^2 - \frac{2(1+\gamma)}{2\alpha + 1}\Delta^{2\alpha+1} + o\left(\Delta^{2\alpha+1}\right).$$

Since

$$I_2' = \int_0^\Delta \left\{x^\alpha - (\Delta - x)^\alpha\right\}^2 e^{-x^\alpha}\, dx$$

$$= \int_0^\Delta \left\{x^\alpha - (\Delta - x)^\alpha\right\}^2 (1 - x^\alpha)\, dx + O\left(\Delta^{3\alpha+1}\right)$$

$$= \int_0^\Delta \left\{x^{2\alpha} - 2x^\alpha(\Delta - x)^\alpha + (\Delta - x)^{2\alpha}\right\} dx + O\left(\Delta^{3\alpha+1}\right)$$

$$= \frac{2\Delta^{2\alpha+1}}{2\alpha + 1} - 2\Delta^{2\alpha+1}\int_0^\Delta \left(\frac{x}{\Delta}\right)^\alpha \left(1 - \frac{x}{\Delta}\right)^\alpha \frac{1}{\Delta}\, dx + O\left(\Delta^{3\alpha+1}\right)$$

$$= \frac{2\Delta^{2\alpha+1}}{2\alpha + 1} - 2\Delta^{2\alpha+1} B(\alpha + 1, \alpha + 1) + O\left(\Delta^{3\alpha+1}\right),$$

we obtain from (6.2.10) and (6.2.11)

$$(6.2.12)\ E_0\left[\Psi_\Delta^2(X)\right] = C(\alpha)\left(I_1' + I_2' + I_3'\right)$$

$$= \frac{\alpha(\alpha - 1)\Gamma\left(1 - \dfrac{1}{\alpha}\right)}{\Gamma\left(\dfrac{1}{\alpha}\right)}\Delta^2$$

$$- \frac{\alpha\left\{B(\alpha + 1, \alpha + 1) + (\gamma/(2\alpha + 1))\right\}}{\Gamma\left(\dfrac{1}{\alpha}\right)}\Delta^{2\alpha+1} + o\left(\Delta^{2\alpha+1}\right).$$

From (6.2.5), we have

$$(6.2.13)\quad E_0\left[\Psi_\Delta^3(X)\right] = \int_{-\infty}^0 \left\{(-x)^\alpha - (\Delta - x)^\alpha\right\}^3 f(x)dx + \int_0^\Delta \left\{x^\alpha - (\Delta - x)^\alpha\right\}^3 f(x)dx$$

$$+ \int_\Delta^\infty \left\{x^\alpha - (x - \Delta)^\alpha\right\}^3 f(x)dx$$

$$= \int_0^\infty \left\{x^\alpha - (x + \Delta)^\alpha\right\}^3 f(x)dx + \int_0^\Delta \left\{x^\alpha - (\Delta - x)^\alpha\right\}^3 f(x)dx$$

$$+ \int_\Delta^\infty \left\{ x^\alpha - (x - \Delta)^\alpha \right\}^3 f(x) dx$$
$$= C(\alpha) (I_1'' + I_2'' + I_3'') \qquad \text{(say)}.$$

Since $h_\Delta(x) = (x + \Delta)^\alpha - x^\alpha$, it follows that

$$(6.2.14) \quad I_1'' + I_3'' = - \int_0^\infty \left\{ (x + \Delta)^\alpha - x^\alpha \right\}^3 e^{-x^\alpha} dx + \int_\Delta^\infty \left\{ x^\alpha - (x - \Delta)^\alpha \right\}^3 e^{-x^\alpha} dx$$

$$= \int_0^\infty \left\{ (x + \Delta)^\alpha - x^\alpha \right\}^3 \left\{ e^{-(x + \Delta)^\alpha} - e^{-x^\alpha} \right\} dx$$

$$= \int_0^\infty h_\Delta^3(x) \left\{ e^{-h_\Delta(x)} - 1 \right\} e^{-x^\alpha} dx.$$

Since

$$\int_0^\infty h_\Delta^4(x) e^{-x^\alpha} dx = \int_0^\infty \left\{ (x + \Delta)^\alpha - x^\alpha \right\}^4 e^{-x^\alpha} dx$$

$$= \alpha^4 \Delta^4 \int_0^\infty x^{4\alpha - 4} e^{-x^\alpha} dx + o\left(\Delta^4 \right) = O\left(\Delta^4 \right),$$

it follows from (6.2.14) that

$$(6.2.15) \qquad\qquad I_1'' + I_3'' = O\left(\Delta^4 \right).$$

We also have

$$(6.2.16) \qquad I_2'' = \int_0^\Delta \left\{ x^\alpha - (\Delta - x)^\alpha \right\}^3 e^{-x^\alpha} dx = O\left(\Delta^{3\alpha+1} \right).$$

From (6.2.13), (6.2.15) and (6.2.16), we obtain

$$(6.2.17) \qquad\qquad E_0 \left[\Psi_\Delta^3(X) \right] = O\left(\Delta^4 \right).$$

Putting $\Delta = t n^{-1/2}$ with $t > 0$, we obtain from (6.2.9), (6.2.12) and (6.2.17)

$$E_0 [Z_n] = -n E_0 \left[\Psi_{t n^{-1/2}}(X) \right]$$

$$= \frac{\alpha(\alpha - 1)\Gamma \left(1 - \dfrac{1}{\alpha} \right)}{2\Gamma \left(\dfrac{1}{\alpha} \right)} t^2 - \frac{2 \left\{ B(\alpha + 1, \alpha + 1) + (\gamma/(2\alpha + 1)) \right\}}{2\Gamma \left(\dfrac{1}{\alpha} \right)}$$

$$\cdot t^{2\alpha+1} n^{-(2\alpha-1)/2} + o\left(n^{-(2\alpha-1)/2} \right)$$

$$= \frac{I}{2} t^2 - \frac{k}{2} t^{2\alpha+1} n^{-(2\alpha-1)/2} + o\left(n^{-(2\alpha-1)/2} \right),$$

$$V_0\left(Z_n\right) = nV_0\left(\Psi_{tn^{-1/2}}(X)\right)$$

$$= \frac{\alpha(\alpha - 1)\Gamma\left(1 - \dfrac{1}{\alpha}\right)}{\Gamma\left(\dfrac{1}{\alpha}\right)}t^2 - \frac{\alpha\left\{B(\alpha + 1, \alpha + 1) + (\gamma/(2\alpha + 1))\right\}}{\Gamma\left(\dfrac{1}{\alpha}\right)}$$

$$\cdot\, t^{2\alpha+1}n^{-(2\alpha-1)/2} + o\left(n^{-(2\alpha-1)/2}\right)$$

$$= It^2 - kt^{2\alpha+1}n^{-(2\alpha-1)/2} + o\left(n^{-(2\alpha-1)/2}\right),$$

$$\mathcal{K}_{3,0}\left(Z_n\right) = -n\mathcal{K}_{3,0}\left(\Psi_{tn^{-1/2}}(X)\right)$$

$$= -nE_0\left[\left\{\Psi_{tn^{-1/2}}(X) - E_0\left[\Psi_{tn^{-1/2}}(X)\right]\right\}^3\right]$$

$$= o\left(n^{-(2\alpha-1)/2}\right),$$

where

$$I = E_\theta\left[\left\{\frac{\partial}{\partial\theta}\log f(X - \theta)\right\}^2\right] = -E_\theta\left[\frac{\partial^2}{\partial\theta^2}\log f(X - \theta)\right]$$

$$= \alpha(\alpha - 1)\Gamma(1 - (1/\alpha))/\Gamma(1/\alpha)$$

and

$$k = \alpha\left\{B(\alpha + 1, \alpha + 1) + \frac{\gamma}{2\alpha + 1}\right\}\Big/\Gamma(1/\alpha).$$

In a similar way to the case under $K : \theta = 0$, we can obtain the asymptotic mean, variance and third-order cumulants under $H : \theta = tn^{-1/2}$. Thus we complete the proof.

Proof of Theorem 6.1.1. Without loss of generality, we assume that $\theta_0 = 0$. We consider the case when $t > 0$. In order to choose α_0 such that

(6.2.18) $$P_{tn^{-1/2}}^n\left\{Z_n \leq \alpha_0\right\} = \frac{1}{2} + o\left(n^{-(2\alpha-1)/2}\right),$$

we have by Lemmas 6.1.2 and 6.1.3

$$\alpha_0 = -\frac{I}{2}t^2 + \frac{k}{2}t^{2\alpha+1}n^{-(2\alpha-1)/2} + o\left(n^{-(2\alpha-1)/2}\right).$$

Since

(6.2.19) $$P_0^n\left\{Z_n \geq \alpha_0\right\} = P_0^n\left\{-\left(Z_n - It^2 - \alpha_0\right) \leq It^2\right\},$$

putting $W_n = -\left(Z_n - It^2 - \alpha_0\right)$, we have from Lemma 6.1.2

(6.2.20) $$E_0\left(W_n\right) = kt^{2\alpha+1}n^{-(2\alpha-1)/2} + o\left(n^{-(2\alpha-1)/2}\right),$$

(6.2.21) $$V_0\left(W_n\right) = It^2 - kt^{2\alpha+1}n^{-(2\alpha-1)/2} + o\left(n^{-(2\alpha-1)/2}\right),$$

(6.2.22) $$\mathcal{K}_{3,0}\left(W_n\right) = o\left(n^{-(2\alpha-1)/2}\right).$$

We obtain by (6.2.19) to (6.2.22) and the Edgeworth expansion

$$(6.2.23) \quad P_0^n \{Z_n \geq \alpha_0\} = P_0^n \{W_n \leq It^2\}$$

$$= \Phi\left(\sqrt{I}t\right) - \frac{k}{2\sqrt{I}}t^{2\alpha}\phi\left(\sqrt{I}t\right)n^{-(2\alpha-1)/2} + o\left(n^{-(2\alpha-1)/2}\right).$$

Here, from (6.2.18), the definition of Z_n and the fundamental lemma of Neyman-Pearson it is noted that a test with the rejection region $\{Z_n \geq \alpha_0\}$ is the most powerful test of level $1/2 + o\left(n^{-(2\alpha-1)/2}\right)$.

Let $\hat{\theta}_n$ be any 2α-th order AMU estimator. Putting $A_{\hat{\theta}_n} = \left\{\sqrt{n}\hat{\theta}_n \leq t\right\}$, we have

$$P_{tn^{-1/2}}^n\left(A_{\hat{\theta}_n}\right) = P_{tn^{-1/2}}^n\left\{\hat{\theta}_n \leq tn^{-1/2}\right\} = \frac{1}{2} + o\left(n^{-(2\alpha-1)/2}\right).$$

Then it is seen that $\chi_{A_{\hat{\theta}_n}}$ of the indicator of $A_{\hat{\theta}_n}$ is a test of level $1/2 + o\left(n^{-(2\alpha-1)/2}\right)$. From (6.2.23), we obtain for any $\hat{\theta}_n \in A_{2\alpha}$

$$P_0^n\left\{\sqrt{n}\hat{\theta}_n \leq t\right\} \leq P_0^n\{W_n \leq It^2\}$$

$$= \Phi\left(\sqrt{I}t\right) - \frac{k}{2\sqrt{I}}t^{2\alpha}\phi\left(\sqrt{I}t\right)n^{-(2\alpha-1)/2} + o\left(n^{-(2\alpha-1)/2}\right),$$

that is,

$$(6.2.24) \qquad P_0^n\left\{\sqrt{In}\hat{\theta}_n \leq t\right\} \leq \Phi(t) - C_0 t^{2\alpha}\phi(t)n^{-(2\alpha-1)/2} + o\left(n^{-(2\alpha-1)/2}\right)$$

for all $t > 0$, where

$$C_0 = \frac{k}{2I^{\alpha+(1/2)}} = \frac{\alpha\{B(\alpha+1,\alpha+1) + (\gamma/(2\alpha+1))\}}{2I^{\alpha+(1/2)}\Gamma(1/\alpha)}.$$

Hence we see that the bound for the 2α-th order asymptotic distribution of 2α-th order AMU estimators for all $t > 0$ is given by (6.2.24). In a similar way to the case $t > 0$, we can obtain the 2α-th order bound for all $t < 0$. Thus we complete the proof.

6.3. THE 2α-TH ORDER ASYMPTOTIC DISTRIBUTION OF THE MAXIMUM LIKELIHOOD ESTIMATOR

In this section we obtain the 2α-th order asymptotic distribution of the maximum likelihood estimator (MLE) and compare it with the 2α-th order asymptotic bound obtained in Section 6.1.

We denote by θ_0 and $\hat{\theta}_{ML}$ the true parameter and the MLE, respectively. It is seen that for real t, $\hat{\theta}_{ML} < \theta_0 + tn^{-1/2}$ if and only if

$$(\partial/\partial\theta)\sum_{i=1}^{n}\log f\left(X_i - \theta_0 - tn^{-1/2}\right) < 0.$$

Without loss of generality, we assume that $\theta_0 = 0$. Hence we see that for each real t

(6.3.1) $\hat{\theta}_{ML} < tn^{-1/2}$ if and only if $\frac{1}{\sqrt{n}} \sum_{i=1}^{n} (d/dx) \log f\left(X_i - tn^{-1/2}\right) > 0.$

Since

$$\frac{d}{dx} \log f(x) = -\alpha |x|^{\alpha-1} \text{sgn} x,$$

we put

(6.3.2) $$U_n = -\frac{1}{\sqrt{n}} \sum_{i=1}^{n} (d/dx) \log f\left(X_i - tn^{-1/2}\right)$$

$$= \frac{\alpha}{\sqrt{n}} \sum_{i=1}^{n} \left| X_i - tn^{-1/2} \right|^{\alpha-1} \text{sgn}\left(X_i - tn^{-1/2}\right).$$

In order to obtain the asymptotic cumulants of U_n, we need the following lemma.

Lemma 6.3.1. If $h_\Delta(x) = (x + \Delta)^\alpha - x^\alpha$ for $\Delta > 0$, then

$$\int_0^\infty x^{\alpha-1} e^{-x^\alpha} h_\Delta(x) dx = \Gamma\left(2 - \frac{1}{\alpha}\right)\Delta + \frac{\alpha-1}{2}\Gamma\left(2 - \frac{2}{\alpha}\right)\Delta^2$$

$$- \frac{\gamma}{2\alpha}\Delta^{2\alpha} + o\left(\Delta^{2\alpha}\right),$$

(6.3.3) $$\int_0^\infty x^{2\alpha-2} e^{-x^\alpha} h_\Delta(x) dx = \Gamma\left(3 - \frac{2}{\alpha}\right)\Delta + \frac{1}{2}(\alpha-1)\Gamma\left(3 - \frac{3}{\alpha}\right)\Delta^2$$

$$+ o\left(\Delta^2\right),$$

(6.3.4) $$\int_0^\infty x^{3\alpha-3} e^{-x^\alpha} h_\Delta(x) dx = \left(3 - \frac{3}{\alpha}\right)\Gamma\left(3 - \frac{3}{\alpha}\right)\Delta + O\left(\Delta^2\right),$$

$$\int_0^\infty x^{\alpha-1} e^{-x^\alpha} h_\Delta^2(x) dx = \alpha\Gamma\left(3 - \frac{2}{\alpha}\right)\Delta^2 + O\left(\Delta^3\right),$$

(6.3.5) $$\int_0^\infty x^{2\alpha-2} e^{-x^\alpha} h_\Delta^2(x) dx = \alpha\Gamma\left(4 - \frac{3}{\alpha}\right)\Delta^2 + O\left(\Delta^3\right),$$

(6.3.6) $$\int_0^\infty x^{3\alpha-3} e^{-x^\alpha} h_\Delta^2(x) dx = O\left(\Delta^2\right),$$

$$\int_0^\infty x^{\alpha-1} e^{-x^\alpha} h_\Delta^3(x) dx = O\left(\Delta^3\right),$$

(6.3.7) $$\int_0^\infty x^{2\alpha-2} e^{-x^\alpha} h_\Delta^3(x) dx = O\left(\Delta^3\right).$$

The proof is given later.

In the following lemma, we obtain the asymptotic mean, variance and third-order cumulant of U_n.

Lemma 6.3.2. *The asymptotic mean, variance and third-order cumulant of* U_n *are given for* $t > 0$ *as follows :*

$$E_0\left(U_n\right) = -It + \frac{k'}{2}t^{2\alpha}n^{-(2\alpha-1)/2} + o\left(n^{-(2\alpha-1)/2}\right),$$

$$V_0\left(U_n\right) = I + o\left(n^{-(2\alpha-1)/2}\right),$$

$$\mathcal{K}_{3,0}\left(U_n\right) = o\left(n^{-(2\alpha-1)/2}\right),$$

where I *is given in Lemma 6.1.2 and*

$$k' = \frac{\alpha^2\left\{B(\alpha, \alpha+1) + (\gamma/\alpha)\right\}}{\Gamma(1/\alpha)}.$$

The proof is given later.

In the following theorem, we obtain the 2α-th order asymptotic distribution of the MLE of θ.

Theorem 6.3.1. *The* 2α-*th order asymptotic distribution of the MLE* $\hat{\theta}_{ML}$ *of* θ *is given by*

$$(6.3.8) \quad P_\theta^n\left\{\sqrt{In}\left(\hat{\theta}_{ML} - \theta\right) \le t\right\} = \Phi(t) - C_1|t|^{2\alpha}\phi(t)n^{-(2\alpha-1)/2}\mathrm{sgn}t + o\left(n^{-(2\alpha-1)/2}\right),$$

where $C_1 = (2\alpha+1)C_0$ *with* $C_0 = \alpha\{B(\alpha+1, \alpha+1) + (\gamma/(2\alpha+1))\}/\{2I^{\alpha+(1/2)}\Gamma(1/\alpha)\}$, *and also the MLE is not* 2α-*th order asymptotically efficient in the sense that its* 2α-*th order asymptotic distribution does not uniformly attain the bound given in Theorem 6.1.1.*

The proof is given later.

Remark 6.3.1. In the double exponential distribution case, that is, the case when $\alpha = 1$, it is shown in Akahira and Takeuchi (1981) that the bound for the second order asymptotic distribution of the second order AMU estimators of θ is given by

$$\Phi(t) - \frac{t^2}{6}\phi(t)n^{-1/2}\mathrm{sgn}t + o\left(n^{-1/2}\right),$$

and the second order asymptotic distribution of the MLE of θ, i.e., the median of $X_1, \ldots, X_n$, is given by

$$(6.3.9) \qquad \Phi(t) - \frac{t^2}{2}\phi(t)n^{-1/2}\mathrm{sgn}t + o\left(n^{-1/2}\right)$$

(see also Section 4.2). The results coincide with the case when $\alpha = 1$ is substituted in the formulae of Theorems 6.1.1 and 6.3.1, but note that the proofs of these theorems do not include the case for $\alpha = 1$ since it does not automatically hold that $\Gamma(\alpha) = (\alpha-1)\Gamma(\alpha-1)$ for $\alpha = 1$.

Here we give the proofs of Lemma 6.3.1, 6.3.2 and Theorem 6.3.1.

Proof of Lemma 6.3.1. From (6.2.2), we have

$$\int_0^\infty x^{\alpha-1}e^{-x^\alpha}h_\Delta(x)dx$$

$$= \int_0^\infty (x+\Delta)^\alpha x^{\alpha-1}e^{-x^\alpha}dx - \int_0^\infty x^{2\alpha-1}e^{-x^\alpha}dx$$

$$= -\frac{1}{\alpha+1}\int_0^\infty (x+\Delta)^{\alpha+1}\left\{(\alpha-1)x^{\alpha-2}-\alpha x^{2\alpha-2}\right\}e^{-x^\alpha}dx - \frac{1}{\alpha}$$

$$= -\frac{1}{\alpha+1}\int_0^\infty \left\{x^{\alpha+1}+(\alpha+1)\Delta x^\alpha+\frac{\alpha(\alpha+1)}{2}\Delta^2 x^{\alpha-1}\right\}$$

$$\cdot\left\{(\alpha-1)x^{\alpha-2}-\alpha x^{2\alpha-2}\right\}e^{-x^\alpha}dx$$

$$-\frac{\alpha-1}{\alpha+1}\int_0^\infty R^*(\Delta)x^{\alpha-2}e^{-x^\alpha}dx - \frac{1}{\alpha}+O\left(\Delta^3\right),$$

where

$$R^*(\Delta) = (x+\Delta)^{\alpha+1} - x^{\alpha+1} - (\alpha+1)\Delta x^\alpha - \frac{1}{2}\alpha(\alpha+1)\Delta^2 x^{\alpha-1}.$$

In a similar way to (6.2.4), we obtain

$$\int_0^\infty R^*(\Delta)x^{\alpha-2}e^{-x^\alpha}dx$$

$$= \frac{1}{2}\alpha(\alpha+1)(\alpha-1)\int_0^\Delta (\Delta-t)^2\left(\int_0^\infty (x+t)^{\alpha-2}x^{\alpha-2}e^{-x^\alpha}dx\right)dt$$

$$= \frac{1}{2}\alpha(\alpha+1)(\alpha-1)B(2\alpha-2,3)B(\alpha-1,3-2\alpha)\Delta^{2\alpha}+o\left(\Delta^{2\alpha}\right)$$

$$= \frac{(\alpha+1)\Gamma(\alpha-1)\Gamma(3-2\alpha)}{4(2\alpha-1)\Gamma(2-\alpha)}\Delta^{2\alpha}+o\left(\Delta^{2\alpha}\right).$$

Hence, we have

$$\int_0^\infty x^{\alpha-1}e^{-x^\alpha}h_\Delta(x)dx$$

$$= -\frac{1}{\alpha+1}\left\{(\alpha-1)\int_0^\infty x^{2\alpha-1}e^{-x^\alpha}dx - \alpha\int_0^\infty x^{3\alpha-1}e^{-x^\alpha}dx\right.$$

$$+(\alpha-1)(\alpha+1)\Delta\int_0^\infty x^{2\alpha-2}e^{-x^\alpha}dx - \alpha(\alpha+1)\Delta\int_0^\infty x^{3\alpha-2}e^{-x^\alpha}dx$$

$$+\frac{1}{2}\alpha(\alpha-1)(\alpha+1)\Delta^2\int_0^\infty x^{2\alpha-3}e^{-x^\alpha}dx$$

$$\left.-\frac{1}{2}\alpha^2(\alpha+1)\Delta^2\int_0^\infty x^{3\alpha-3}e^{-x^\alpha}dx\right\}$$

$$-\frac{1}{\alpha}-\frac{(\alpha-1)\Gamma(\alpha-1)\Gamma(3-2\alpha)}{4(2\alpha-1)\Gamma(2-\alpha)}\Delta^{2\alpha}+o\left(\Delta^{2\alpha}\right)$$

$$= -\frac{1}{\alpha+1}\left\{\frac{\alpha-1}{\alpha} - 2 + \frac{1}{\alpha}(\alpha-1)(\alpha+1)\Gamma\left(2-\frac{1}{\alpha}\right)\Delta - (\alpha+1)\Gamma\left(3-\frac{1}{\alpha}\right)\Delta \right.$$

$$\left. + \frac{1}{2}(\alpha-1)(\alpha+1)\Gamma\left(2-\frac{2}{\alpha}\right)\Delta^2 - \frac{1}{2}\alpha(\alpha+1)\Gamma\left(3-\frac{2}{\alpha}\right)\Delta^2\right\}$$

$$- \frac{1}{\alpha} - \frac{(\alpha-1)\Gamma(\alpha-1)\Gamma(3-2\alpha)}{4(2\alpha-1)\Gamma(2-\alpha)}\Delta^{2\alpha} + O\left(\Delta^3\right)$$

$$= \Gamma\left(2-\frac{1}{\alpha}\right)\Delta + \frac{\alpha-1}{2}\Gamma\left(2-\frac{2}{\alpha}\right)\Delta^2 - \frac{\gamma}{2\alpha}\Delta^{2\alpha} + o\left(\Delta^{2\alpha}\right).$$

In a similar way to the above, we can obtain (6.3.3) and (6.3.4). We also have

$$\int_0^\infty x^{\alpha-1}e^{-x^\alpha}h_\Delta^2(x)dx$$

$$= \int_0^\infty (x+\Delta)^{2\alpha}x^{\alpha-1}e^{-x^\alpha}dx - 2\int_0^\infty (x+\Delta)^\alpha x^{2\alpha-1}e^{-x^\alpha}dx + \int_0^\infty x^{3\alpha-1}e^{-x^\alpha}dx$$

$$= \int_0^\infty \left\{x^{2\alpha} + 2\alpha x^{2\alpha-1}\Delta + \alpha(2\alpha-1)x^{2\alpha-2}\Delta^2\right\}x^{\alpha-1}e^{-x^\alpha}dx$$

$$- 2\int_0^\infty \left\{x^\alpha + \alpha x^{\alpha-1}\Delta + \frac{1}{2}\alpha(\alpha-1)x^{\alpha-2}\Delta^2\right\}x^{2\alpha-1}e^{-x^\alpha}dx$$

$$+ \frac{2}{\alpha} + O\left(\Delta^3\right)$$

$$= \alpha\Gamma\left(3-\frac{2}{\alpha}\right)\Delta^2 + O\left(\Delta^3\right),$$

and similarly get (6.3.5) and (6.3.6). From (6.2.2), we obtain

$$\int_0^\infty x^{\alpha-1}e^{-x^\alpha}h_\Delta^3(x)dx = \int_0^\infty x^{\alpha-1}e^{-x^\alpha}\left(\alpha x^{\alpha-1}\Delta\right)^3 dx + o\left(\Delta^3\right)$$

$$= \alpha^2\Gamma\left(4-\frac{3}{\alpha}\right)\Delta^3 + o\left(\Delta^3\right)$$

$$= O\left(\Delta^3\right),$$

and similarly have (6.3.7). Thus we complete the proof.

Proof of Lemma 6.3.2. Putting $\Delta = tn^{-1/2}$, we obtain

(6.3.10) $E_0\left[|X-\Delta|^{\alpha-1}\mathrm{sgn}(X-\Delta)\right]$

$$= C(\alpha)\left\{\int_\Delta^\infty (x-\Delta)^{\alpha-1}e^{-x^\alpha}dx - \int_0^\Delta (\Delta-x)^{\alpha-1}e^{-x^\alpha}dx\right.$$

$$\left. - \int_{-\infty}^0 (\Delta-x)^{\alpha-1}e^{-|x|^\alpha}dx\right\}$$

$$= C(\alpha)\left\{\int_0^\infty x^{\alpha-1}e^{-(x+\Delta)^\alpha}dx - \int_0^\infty (x+\Delta)^{\alpha-1}e^{-x^\alpha}dx\right.$$

$$-\int_0^\Delta (\Delta - x)^{\alpha-1} e^{-x^\alpha} dx \Big\}$$

$$= C(\alpha) \bigg[\int_0^\infty x^{\alpha-1} \left\{ e^{-(x+\Delta)^\alpha} - e^{-x^\alpha} \right\} dx - \int_0^\infty \left\{ (x+\Delta)^{\alpha-1} - x^{\alpha-1} \right\} e^{-x^\alpha} dx$$

$$-\int_0^\Delta (\Delta - x)^{\alpha-1} e^{-x^\alpha} dx \bigg]$$

$$= C(\alpha)(J_1 + J_2 + J_3) \qquad \text{(say)}.$$

Putting $h_\Delta(x) = (x+\Delta)^\alpha - x^\alpha$, we have from Lemma 6.3.1

(6.3.11) $\quad J_1 = \int_0^\infty x^{\alpha-1} \left\{ e^{-(x+\Delta)^\alpha} - e^{-x^\alpha} \right\} dx$

$$= \int_0^\infty x^{\alpha-1} e^{-x^\alpha} \left\{ e^{-h_\Delta(x)} - 1 \right\} dx$$

$$= -\int_0^\infty x^{\alpha-1} e^{-x^\alpha} h_\Delta(x) dx + \frac{1}{2} \int_0^\infty x^{\alpha-1} e^{-x^\alpha} h_\Delta^2(x) dx + O\left(\Delta^3\right)$$

$$= -\Gamma\left(2 - \frac{1}{\alpha}\right)\Delta - \frac{\alpha-1}{2}\Gamma\left(2 - \frac{2}{\alpha}\right)\Delta^2 + \frac{\alpha}{2}\Gamma\left(3 - \frac{2}{\alpha}\right)\Delta^2$$

$$+ \frac{\gamma}{2\alpha}\Delta^{2\alpha} + o\left(\Delta^{2\alpha}\right)$$

$$= -\left(1 - \frac{1}{\alpha}\right)\Gamma\left(1 - \frac{1}{\alpha}\right)\Delta + \frac{\alpha-1}{2}\Gamma\left(2 - \frac{2}{\alpha}\right)\Delta^2$$

$$+ \frac{\gamma}{2\alpha}\Delta^{2\alpha} + o\left(\Delta^{2\alpha}\right).$$

From (6.2.2), we obtain

(6.3.12) $\quad -J_2 = \int_0^\infty (x+\Delta)^{\alpha-1} e^{-x^\alpha} dx - \int_0^\infty x^{\alpha-1} e^{-x^\alpha} dx$

$$= -\frac{\Delta^\alpha}{\alpha} + \int_0^\infty (x+\Delta)^\alpha x^{\alpha-1} e^{-x^\alpha} dx - \frac{1}{\alpha}.$$

Since by a similar way to (6.2.4)

$$\int_0^\infty (x+\Delta)^\alpha x^{\alpha-1} e^{-x^\alpha} dx$$

$$= \frac{1}{\alpha} + \Gamma\left(2 - \frac{1}{\alpha}\right)\Delta + \frac{\alpha-1}{2}\Gamma\left(2 - \frac{2}{\alpha}\right)\Delta^2 - \frac{\gamma}{2\alpha}\Delta^{2\alpha} + o\left(\Delta^{2\alpha}\right),$$

it follows from (6.3.12) that

(6.3.13) $\quad J_2 = \frac{\Delta^\alpha}{\alpha} - \Gamma\left(2 - \frac{1}{\alpha}\right)\Delta - \frac{\alpha}{2}\left(1 - \frac{1}{\alpha}\right)\Gamma\left(2 - \frac{2}{\alpha}\right)\Delta^2$

$$+ \frac{\gamma}{2\alpha}\Delta^{2\alpha} + o\left(\Delta^{2\alpha}\right).$$

We also have

$$(6.3.14) \qquad -J_3 = \int_0^\Delta (\Delta - x)^{\alpha-1} e^{-x^\alpha} dx$$

$$= \int_0^\Delta (\Delta - x)^{\alpha-1} (1 - x^\alpha) \, dx + O\left(\Delta^{3\alpha}\right)$$

$$= \frac{\Delta^\alpha}{\alpha} - \Delta^{2\alpha} \int_0^1 \left(1 - \frac{x}{\Delta}\right)^{\alpha-1} \left(\frac{x}{\Delta}\right)^\alpha \frac{1}{\Delta} dx + O\left(\Delta^{3\alpha}\right)$$

$$= \frac{\Delta^\alpha}{\alpha} - B(\alpha, \alpha+1) \Delta^{2\alpha} + O\left(\Delta^{3\alpha}\right).$$

From (6.3.10), (6.3.11), (6.3.13) and (6.3.14), we obtain

$$(6.3.15) \quad E_0 \left[|X - \Delta|^{\alpha-1} \mathrm{sgn}(X - \Delta)\right]$$

$$= C(\alpha) \left\{ -\frac{2(\alpha-1)}{\alpha} \Gamma\left(\frac{\alpha-1}{\alpha}\right) \Delta + \left(B(\alpha, \alpha+1) + \frac{\gamma}{\alpha}\right) \Delta^{2\alpha} \right\} + o\left(\Delta^{2\alpha}\right)$$

$$= -(\alpha-1) \frac{\Gamma(1 - (1/\alpha))}{\Gamma(1/\alpha)} \Delta + \frac{\alpha\{B(\alpha, \alpha+1) + (\gamma/\alpha)\}}{2\Gamma(1/\alpha)} \Delta^{2\alpha} + o\left(\Delta^{2\alpha}\right).$$

Next we have

$$(6.3.16) \qquad E_0 \left[|X - \Delta|^{2\alpha-2}\right]$$

$$= C(\alpha) \left\{ \int_\Delta^\infty (x - \Delta)^{2\alpha-2} e^{-x^\alpha} dx + \int_{-\infty}^\Delta (\Delta - x)^{2\alpha-2} e^{-|x|^\alpha} dx \right\}$$

$$= C(\alpha) \left\{ \int_0^\infty x^{2\alpha-2} e^{-(x+\Delta)^\alpha} dx + \int_0^\infty (x + \Delta)^{2\alpha-2} e^{-x^\alpha} dx \right.$$

$$\left. + \int_0^\Delta (\Delta - x)^{2\alpha-2} e^{-x^\alpha} dx \right\}$$

$$= C(\alpha) \left[2 \int_0^\infty x^{2\alpha-2} e^{-x^\alpha} dx + \int_0^\infty \left\{ e^{-(x+\Delta)^\alpha} - e^{-x^\alpha} \right\} x^{2\alpha-2} dx \right.$$

$$\left. + \int_0^\infty \left\{ (x + \Delta)^{2\alpha-2} - x^{2\alpha-2} \right\} e^{-x^\alpha} dx + \int_0^\Delta (\Delta - x)^{2\alpha-2} e^{-x^\alpha} dx \right]$$

$$= C(\alpha) \left[J_1' + J_2' + J_3' + J_4' \right] \qquad \text{(say)}.$$

From (6.2.2), we obtain

$$(6.3.17) \qquad J_1' = 2 \int_0^\infty x^{2\alpha-2} e^{-x^\alpha} dx = \frac{2}{\alpha} \Gamma\left(2 - \frac{1}{\alpha}\right) = \frac{2}{\alpha} \left(1 - \frac{1}{\alpha}\right) \Gamma\left(1 - \frac{1}{\alpha}\right).$$

Since

$$
\begin{aligned}
J_2' &= \int_0^\infty \left\{ e^{-(x+\Delta)^\alpha} - e^{-x^\alpha} \right\} x^{2\alpha-2} dx \\
&= \int_0^\infty x^{2\alpha-2} e^{-x^\alpha} \left\{ e^{-h_\Delta(x)} - 1 \right\} dx \\
&= -\int_0^\infty x^{2\alpha-2} e^{-x^\alpha} h_\Delta(x) dx + \frac{1}{2} \int_0^\infty x^{2\alpha-2} e^{-x^\alpha} h_\Delta^2(x) dx \\
&\quad - \frac{1}{6} \int_0^\infty x^{2\alpha-2} e^{-x^\alpha} h_\Delta^3(x) dx + O\left(\Delta^3\right),
\end{aligned}
$$

it follows from Lemma 6.3.1 that

$$
(6.3.18) \qquad J_2' = -\Gamma\left(3 - \frac{2}{\alpha}\right)\Delta + (\alpha-1)\Gamma\left(3 - \frac{3}{\alpha}\right)\Delta^2 + O\left(\Delta^3\right).
$$

From (6.2.2), we have

$$
\begin{aligned}
(6.3.19) \quad J_3' &= \int_0^\infty \left\{ (x+\Delta)^{2\alpha-2} - x^{2\alpha-2} \right\} e^{-x^\alpha} dx \\
&= -\frac{1}{2\alpha-1}\Delta^{2\alpha-1} + \frac{\alpha}{2\alpha-1}\int_0^\infty (x+\Delta)^{2\alpha-1} x^{\alpha-1} e^{-x^\alpha} dx - \frac{1}{\alpha}\Gamma\left(\frac{2\alpha-1}{\alpha}\right) \\
&= -\frac{1}{2\alpha-1}\Delta^{2\alpha-1} - \frac{1}{\alpha}\Gamma\left(\frac{2\alpha-1}{\alpha}\right) \\
&\quad + \frac{\alpha}{2\alpha-1}\left\{ \int_0^\infty x^{3\alpha-2} e^{-x^\alpha} dx + (2\alpha-1)\Delta \int_0^\infty x^{3\alpha-3} e^{-x^\alpha} dx \right. \\
&\qquad \left. + (\alpha-1)(2\alpha-1)\Delta^2 \int_0^\infty x^{3\alpha-4} e^{-x^\alpha} dx \right\} + o\left(\Delta^2\right) \\
&= -\frac{1}{2\alpha-1}\Delta^{2\alpha-1} - \frac{1}{\alpha}\Gamma\left(2 - \frac{1}{\alpha}\right) + \frac{1}{2\alpha-1}\Gamma\left(3 - \frac{1}{\alpha}\right) \\
&\quad + \Gamma\left(3 - \frac{2}{\alpha}\right)\Delta + (\alpha-1)\Gamma\left(3 - \frac{3}{\alpha}\right)\Delta^2 + O\left(\Delta^3\right) \\
&= -\frac{1}{2\alpha-1}\Delta^{2\alpha-1} + \Gamma\left(3 - \frac{2}{\alpha}\right)\Delta + (\alpha-1)\Gamma\left(3 - \frac{3}{\alpha}\right)\Delta^2 + o\left(\Delta^2\right).
\end{aligned}
$$

We also obtain

$$
\begin{aligned}
(6.3.20) \quad J_4' &= \int_0^\Delta (\Delta - x)^{2\alpha-2} e^{-x^\alpha} dx \\
&= \int_0^\Delta (\Delta - x)^{2\alpha-2} dx - \int_0^\Delta x^\alpha (\Delta - x)^{2\alpha-2} dx + O\left(\Delta^{4\alpha-1}\right) \\
&= \frac{\Delta^{2\alpha-1}}{2\alpha-1} - \Delta^{3\alpha-1}\int_0^\Delta \left(\frac{x}{\Delta}\right)^\alpha \left(1 - \frac{x}{\Delta}\right)^{2\alpha-2} \frac{1}{\Delta} dx + O\left(\Delta^{4\alpha-1}\right) \\
&= \frac{\Delta^{2\alpha-1}}{2\alpha-1} - B(\alpha+1, 2\alpha-1)\Delta^{3\alpha-1} + O\left(\Delta^{4\alpha-1}\right).
\end{aligned}
$$

From (6.3.16) to (6.3.20), we have

$$E_0\left[|X - \Delta|^{2\alpha-2}\right] = \frac{\alpha-1}{\alpha\Gamma(1/\alpha)}\Gamma\left(1 - \frac{1}{\alpha}\right) + \frac{\alpha(\alpha-1)}{\Gamma(1/\alpha)}\Gamma\left(3 - \frac{3}{\alpha}\right)\Delta^2 + o\left(\Delta^2\right),$$

hence, by (6.3.15)

$$(6.3.21)\quad V_0\left(|X - \Delta|^{\alpha-1}\mathrm{sgn}(X - \Delta)\right) = \frac{\alpha-1}{\alpha\Gamma(1/\alpha)}\Gamma\left(1 - \frac{1}{\alpha}\right)$$

$$+ (\alpha-1)\left[\frac{\alpha}{\Gamma(1/\alpha)}\Gamma\left(3 - \frac{3}{\alpha}\right)\right.$$

$$\left. - (\alpha-1)\left\{\frac{\Gamma(1 - (1/\alpha))}{\Gamma(1/\alpha)}\right\}^2\right]\Delta^2 + o\left(\Delta^2\right).$$

Third, we have

$$(6.3.22)\qquad E_0\left[|X - \Delta|^{3\alpha-3}\mathrm{sgn}(X - \Delta)\right]$$

$$= C(\alpha)\left\{\int_\Delta^\infty (x - \Delta)^{3\alpha-3}e^{-x^\alpha}dx - \int_0^\Delta (\Delta - x)^{3\alpha-3}e^{-x^\alpha}dx\right.$$

$$\left. - \int_{-\infty}^0 (\Delta - x)^{3\alpha-3}e^{-|x|^\alpha}dx\right\}$$

$$= C(\alpha)\left[\int_0^\infty x^{3\alpha-3}\left\{e^{-(x+\Delta)^\alpha} - e^{-x^\alpha}\right\}dx\right.$$

$$- \int_0^\infty \left\{(x + \Delta)^{3\alpha-3} - x^{3\alpha-3}\right\}e^{-x^\alpha}dx$$

$$\left. - \int_0^\Delta (\Delta - x)^{3\alpha-3}e^{-x^\alpha}dx\right]$$

$$= C(\alpha)\left(J_1'' + J_2'' + J_3''\right)\qquad (\text{say}).$$

Since

$$J_1'' = \int_0^\infty x^{3\alpha-3}\left\{e^{-(x+\Delta)^\alpha} - e^{-x^\alpha}\right\}dx = \int_0^\infty x^{3\alpha-3}e^{-x^\alpha}\left\{e^{-h_\Delta(x)} - 1\right\}dx$$

$$= -\int_0^\infty x^{3\alpha-3}e^{-x^\alpha}h_\Delta(x)dx + \frac{1}{2}\int_0^\infty x^{3\alpha-3}e^{-x^\alpha}h_\Delta^2(x)dx + o\left(\Delta^2\right),$$

it follows by Lemma 6.3.1 that

$$(6.3.23)\qquad J_1'' = -3\left(1 - \frac{1}{\alpha}\right)\Gamma\left(3 - \frac{3}{\alpha}\right)\Delta + O\left(\Delta^2\right).$$

From (6.2.2), we have

$$(6.3.24)\qquad -J_2'' = \int_0^\infty \left\{(x + \Delta)^{3\alpha-3} - x^{3\alpha-3}\right\}e^{-x^\alpha}dx$$

$$= 3(\alpha - 1)\Delta \int_0^\infty x^{3\alpha-4} e^{-x^\alpha} dx + o(\Delta)$$

$$= 3\left(1 - \frac{1}{\alpha}\right) \Gamma\left(3 - \frac{3}{\alpha}\right) \Delta + o(\Delta),$$

and also

$$(6.3.25) \quad -J_3'' = \int_0^\Delta (\Delta - x)^{3\alpha-3} e^{-x^\alpha} dx$$

$$= \frac{\Delta^{3\alpha-2}}{3\alpha - 2} - \Delta^{4\alpha-2} \int_0^\Delta \left(\frac{x}{\Delta}\right)^\alpha \left(1 - \frac{x}{\Delta}\right)^{3\alpha-3} \frac{1}{\Delta} dx + O\left(\Delta^{5\alpha-2}\right)$$

$$= \frac{\Delta^{3\alpha-2}}{3\alpha - 2} + O\left(\Delta^{4\alpha-2}\right).$$

From (6.3.22) to (6.3.25), we obtain

$$E_0\left[|X - \Delta|^{3\alpha-3}\text{sgn}(X - \Delta)\right] = -\frac{3(\alpha - 1)}{\Gamma(1/\alpha)} \Gamma\left(3 - \frac{3}{\alpha}\right) \Delta + o(\Delta),$$

hence

$$(6.3.26) \quad \mathcal{K}_{3,0}\left(|X - \Delta|^{\alpha-1}\text{sgn}(X - \Delta)\right)$$

$$= E_0\left[\left\{|X - \Delta|^{\alpha-1}\text{sgn}(X - \Delta) - E_0\left[|X - \Delta|^{\alpha-1}\text{sgn}(X - \Delta)\right]\right\}^3\right]$$

$$= O(\Delta).$$

Putting $\Delta = tn^{-1/2}$ with $t > 0$, we have from (6.3.2), (6.3.15), (6.3.21) and (6.3.26)

$$E_0(U_n) = \alpha\sqrt{n} E_0\left[\left|X - tn^{-1/2}\right|^{\alpha-1} \text{sgn}\left(X - tn^{-1/2}\right)\right]$$

$$= -\alpha(\alpha - 1)\frac{\Gamma(1 - (1/\alpha))}{\Gamma(1/\alpha)} t$$

$$+ \frac{\alpha^2\{B(\alpha, \alpha + 1) + (\gamma/\alpha)\}}{2\Gamma(1/\alpha)} t^{2\alpha} n^{-(2\alpha-1)/2} + o\left(n^{-(2\alpha-1)/2}\right)$$

$$= -It + \frac{k'}{2} t^{2\alpha} n^{-(2\alpha-1)/2} + o\left(n^{-(2\alpha-1)/2}\right),$$

$$V_0(U_n) = \alpha^2 V_0\left(\left|X - tn^{-1/2}\right|^{\alpha-1} \text{sgn}\left(X - tn^{-1/2}\right)\right)$$

$$= \alpha(\alpha - 1)\frac{\Gamma(1 - (1/\alpha))}{\Gamma(1/\alpha)} + o\left(n^{-(2\alpha-1)/2}\right)$$

$$= I + o\left(n^{-(2\alpha-1)/2}\right),$$

$$\mathcal{K}_{3,0}(U_n) = \frac{\alpha^3}{\sqrt{n}} \mathcal{K}_{3,0}\left(\left|X - tn^{-1/2}\right|^{\alpha-1} \text{sgn}\left(X - tn^{-1/2}\right)\right)$$

$$= o\left(n^{-(2\alpha-1)/2}\right),$$

where $I = \alpha(\alpha - 1)\Gamma(1 - (1/\alpha))/\Gamma(1/\alpha)$ and $k' = \alpha^2\{B(\alpha, \alpha + 1) + (\gamma/\alpha)\}/\Gamma(1/\alpha)$. This completes the proof.

Proof of Theorem 6.3.1. Since the density $f(x) = C(\alpha)e^{-|x|^\alpha}$ is symmetric about the origin, we see that the MLE of θ is a 2α-th order AMU. We consider the case when $t > 0$. Using the Edgeworth expansion, we have by (6.3.1), (6.3.2) and Lemma 6.3.2

$$P_\theta^n\left\{\sqrt{n}\left(\hat{\theta}_{ML} - \theta\right) \le t\right\}$$

$$= P_\theta^n\left\{U_n \le 0\right\}$$

$$= \Phi\left(\sqrt{I}t\right) - \frac{k'}{2\sqrt{I}}t^{2\alpha}\phi\left(\sqrt{I}t\right)n^{-(2\alpha-1)/2} + o\left(n^{-(2\alpha-1)/2}\right),$$

that is,

(6.3.27) $P_\theta^n\left\{\sqrt{In}\left(\hat{\theta}_{ML} - \theta\right) \le t\right\}$

$$= \Phi(t) - \frac{k'}{2I^{\alpha+(1/2)}}t^{2\alpha}\phi(t)n^{-(2\alpha-1)/2} + o\left(n^{-(2\alpha-1)/2}\right)$$

$$= \Phi(t) - C_1 t^{2\alpha}\phi(t)n^{-(2\alpha-1)/2} + o\left(n^{-(2\alpha-1)/2}\right),$$

where $C_1 = k'/\left\{2I^{\alpha+(1/2)}\right\}$.

In a similar way to the case $t > 0$, we obtain for $t < 0$

(6.3.28) $P_\theta^n\left\{\sqrt{In}\left(\hat{\theta}_{ML} - \theta\right) \le t\right\}$

$$= \Phi(t) + C_1|t|^{2\alpha}\phi(t)n^{-(2\alpha-1)/2} + o\left(n^{-(2\alpha-1)/2}\right).$$

Hence (6.3.27) and (6.3.28) imply (6.3.8). Since $k' = \alpha^2\{B(\alpha, \alpha + 1) + (\gamma/\alpha)\}/\Gamma(1/\alpha)$ and $C_0 = \alpha\{B(\alpha + 1, \alpha + 1) + (\gamma/(2\alpha + 1))\}/\left\{2I^{\alpha+(1/2)}\Gamma(1/\alpha)\right\}$, it is seen that $C_1 = (2\alpha + 1)C_0$. Since $C_1 > C_0$ for $1 < \alpha < 3/2$, it follows from Theorem 6.1.1 and (6.3.8) that the MLE is not 2α-th order asymptotically efficient in the sense that its 2α-th order asymptotic distribution does not uniformly attain the bound given in Theorem 6.1.1. This completes the proof.

6.4. THE AMOUNT OF THE LOSS OF ASYMPTOTIC INFORMATION OF THE MAXIMUM LIKELIHOOD ESTIMATOR

In the section we obtain the amount of the loss of asymptotic information of the MLE $\hat{\theta}_{ML}$ using its second order asymptotic distribution (6.3.8). Differentiating the right-hand side of (6.3.8) w.r.t. t, we have the second order asymptotic density $g(t)$ of $\sqrt{In}(\hat{\theta}_{ML} - \theta)$ as follows :

(6.4.1) $g(t) = \phi(t)\left\{1 - C_1\left(2\alpha|t|^{2\alpha-1} - |t|^{2\alpha+1}\right)n^{-(2\alpha-1)/2}\right\} + o\left(n^{-(2\alpha-1)/2}\right)$

$$\text{for} \quad -\infty < t < \infty.$$

In general, we obtain for $\alpha > 0$

$$(6.4.2) \quad \int_{-\infty}^{\infty} |t|^{\alpha} \phi(t) dt = \frac{2}{\sqrt{2\pi}} \int_{0}^{\infty} t^{\alpha} e^{-t^2/2} dt$$

$$= \frac{2^{\alpha/2}}{\sqrt{\pi}} \int_{0}^{\infty} u^{\{(\alpha+1)/2\}-1} e^{-u} du$$

(after transformation $u = t^2/2$)

$$= \frac{2^{\alpha/2}}{\sqrt{\pi}} \Gamma\left(\frac{\alpha+1}{2}\right).$$

Since, for sufficiently large n,

$$\frac{d}{dt} \log g(t) = -t - C_1 \left\{ 2\alpha(2\alpha - 1)|t|^{2\alpha-2} - (2\alpha + 1)|t|^{2\alpha} \right\}$$

$$\cdot n^{-(2\alpha-1)/2} \mathrm{sgn}t + o\left(n^{-(2\alpha-1)/2}\right),$$

it follows from (6.4.1) and (6.4.2) that the asymptotic information amount I_{ML} of the MLE is given by

$$I_{ML} = nI \int_{-\infty}^{\infty} \left\{ \frac{d}{dt} \log g(t) \right\}^2 g(t) dt$$

$$= nI \int_{-\infty}^{\infty} \phi(t) \Big[t^2 + C_1 \{ 4\alpha(2\alpha - 1)|t|^{2\alpha-1}$$

$$- 2(3\alpha + 1)|t|^{2\alpha+1} + |t|^{2\alpha+3} \} n^{-(2\alpha-1)/2} \Big] dt + o\left(n^{(3/2)-\alpha}\right)$$

$$= nI \left\{ 1 - \frac{2^{\alpha+(3/2)}}{\sqrt{\pi}} C_1 \Gamma(\alpha + 1) n^{-(2\alpha-1)/2} \right\} + o\left(n^{(3/2)-\alpha}\right).$$

Hence, the amount of the loss L of asymptotic information of the MLE is given by

$$L = nI - I_{ML} = \frac{2^{\alpha+(3/2)}}{\sqrt{\pi}} C_1 I \Gamma(\alpha + 1) n^{(3/2)-\alpha} + o\left(n^{(3/2)-\alpha}\right)$$

$$= \frac{2^{\alpha+(1/2)} \alpha(2\alpha + 1) \Gamma(\alpha + 1) \{ B(\alpha + 1, \alpha + 1) + (\gamma/(2\alpha + 1)) \}}{\sqrt{\pi} I^{\alpha-(1/2)} \Gamma(1/\alpha)} n^{(3/2)-\alpha}$$

$$+ o\left(n^{(3/2)-\alpha}\right).$$

In a similar way to the above, it follows from (6.3.9) that, in the double-exponential distribution case, namely, when $\alpha = 1$, the amount of the loss of asymptotic information of the MLE is given by $2\sqrt{2n/\pi} + o(\sqrt{n})$, since $I = 1$. The amount is consistent with the value given by (4.2.10).

CHAPTER 7

"3/2-TH" AND SECOND ORDER ASYMPTOTICS OF THE GENERALIZED BAYES ESTIMATORS FOR A FAMILY OF TRUNCATED DISTRIBUTIONS

In this chapter we consider the 3/2-th and second order asymptotic optimality of estimators for a family of truncated distributions. Then we obtain the asymptotic density of the generalized Bayes estimator up to the second order, i.e., the order n^{-1}, and have the upper bound for the concentration probability around the true parameter for second order asymptotically median unbiased (AMU) estimators, up to the second order, i.e., the order n^{-1}, in the cases of symmetrically truncated densities, which may not always uniformly attained in the class of all second order AMU estimators. Some examples are given. We also see that the sum and the difference of unilateral differential coefficients of the density at extreme points of its support, in addition to the Fisher information, play an important part in the second order asymptotic behavior of estimators.

7.1. DEFINITIONS AND ASSUMPTIONS

Let $\mathcal{X}$ be an abstract sample space whose generic point is denoted by x, $\mathcal{B}$ a σ-field of subsets of $\mathcal{X}$ and $\{P_\theta : \theta \in \Theta\}$ a set of probability measures on $\mathcal{B}$, where Θ is called a parameter space. We assume that Θ is an open subset of Euclidean 1-space $\mathbf{R}^1$. We denote by $(\mathcal{X}^{(n)}, \mathcal{B}^{(n)})$ the n-fold direct products of $(\mathcal{X}, \mathcal{B})$. Consider n-fold product measures P_θ^n of P_θ. An estimator of θ is defined to be a sequence $\{\hat{\theta}_n\}$ of $\mathcal{B}^{(n)}$-measurable functions $\hat{\theta}_n$ on $\mathcal{X}^{(n)}$ into Θ. For simplicity we denote $\{\hat{\theta}_n\}$ by $\hat{\theta}_n$.

For an increasing sequence of positive numbers $\{c_n\}$ (c_n tending to infinity) an estimator $\hat{\theta}_n$ is called $\{c_n\}$-consistent if for any $\eta \in \Theta$ there exists a sufficiently small positive number δ such that

$$\lim_{L \to \infty} \overline{\lim_{n \to \infty}} \sup_{\theta \,:\, |\theta - \eta| < \delta} P_\theta^n \left\{ c_n \left| \hat{\theta}_n - \theta \right| \geq L \right\} = 0.$$

For any $k \geq 1$, a $\{c_n\}$-consistent estimator $\hat{\theta}_n$ is said to be k-th order asymptotically median unbiased (k-th order AMU for short) if for any $\eta \in \Theta$ there exists a positive number δ such that

$$\lim_{n \to \infty} \sup_{\theta:|\theta-\eta|<\delta} c_n^{k-1} \left| P_\theta^n \left\{ \hat{\theta}_n \leq \theta \right\} - \frac{1}{2} \right| = 0 \ ;$$

$$\lim_{n \to \infty} \sup_{\theta:|\theta-\eta|<\delta} c_n^{k-1} \left| P_\theta^n \left\{ \hat{\theta}_n \geq \theta \right\} - \frac{1}{2} \right| = 0.$$

For a k-th order AMU estimator $\hat{\theta}_n^*$ it is called k-th order two-sided asymptotically efficient if for any k-th order AMU estimator $\hat{\theta}_n$, any $\theta \in \Theta$ and any $t > 0$

$$\varlimsup_{n \to \infty} c_n^{k-1} \left[P_\theta^n \left\{ c_n \left| \hat{\theta}_n^* - \theta \right| \leq t \right\} - P_\theta^n \left\{ c_n \left| \hat{\theta}_n - \theta \right| \leq t \right\} \right] \geq 0.$$

Let $\mathcal{X} = \Theta = \mathbf{R}^1$, and we suppose that for each $\theta \in \Theta$, P_θ is absolutely continuous with respect to a σ-finite measure μ and constitutes a location parameter family. Then we denote the density $dP_\theta/d\mu(x)$ by $f(x, \theta)$ and $f(x, \theta) = f_0(x - \theta)$. Let $X_1, X_2, \ldots, X_n, \ldots$ be a sequence of independent and identically distributed random variables with the density $f_0(x - \theta)$. We assume the following conditions :
(A.7.1.1) $f_0(x) > 0$ for $a < x < b$;
    $f_0(x) = 0$ for $x \leq a$, $x \geq b$.
(A.7.1.2) $f_0(x)$ is twice continuously differentiable in the open interval (a, b) and

$$\lim_{x \to a+0} f_0(x) = \lim_{x \to b-0} f_0(x) = c,$$

where c is some positive constant.

(A.7.1.3)
$$I = \int_a^b \frac{\{f_0'(x)\}^2}{f_0(x)} d\mu(x) < \infty.$$

In this situation, it is known that the order c_n of consistency is equal to n (see Section 3.5). First we have

$$\prod_{i=1}^n f_0(x_i - \theta) > 0 \qquad \text{for} \quad \underline{\theta} < \theta < \overline{\theta},$$

$$\prod_{i=1}^n f_0(x_i - \theta) = 0 \qquad \text{otherwise},$$

where $\underline{\theta} = \max_{1 \leq i \leq n} x_i - b$ and $\overline{\theta} = \min_{1 \leq i \leq n} x_i - a$. Let $L(u)$ be a three times continuously differentiable nonnegative and monotone increasing function of $|u|$. Then a generalized Bayes estimator, with respect to the loss L and the Lebesgue measure, is given by $\hat{\theta}$ which minimizes

$$\int_{\underline{\theta}}^{\overline{\theta}} L \left(\hat{\theta} - \theta \right) \prod_{j=1}^n f_0 \left(x_j - \theta \right) d\theta$$

for almost all $(x_1, \ldots, x_n)$. We put

$$Z_1 = -\frac{1}{\sqrt{n}} \sum_{j=1}^{n} \ell^{(1)}(X_j); \quad Z_2 = \frac{1}{\sqrt{n}} \sum_{j=1}^{n} \ell^{(2)}(X_j) + \sqrt{n} I,$$

where $\ell^{(j)} = (d^j/dx^j) \ell(x)$ $(j = 1, 2)$ with $\ell(x) = \log f_0(x)$ for $a < x < b$. We assume that $(d/du)L(0) = 0$.

7.2. GENERALIZED BAYES ESTIMATORS FOR A FAMILY OF TRUNCATED DISTRIBUTIONS

In this section we obtain the asymptotic density of the generalized Bayes estimator up to the order $o(n^{-1})$ and the concentration probability of the estimator. Without loss of generality we assume that $\theta = 0$ in the previous formulation.

Theorem 7.2.1. *Under the conditions (A.7.1.1) to (A.7.1.3), the generalized Bayes estimator $\hat{\theta}_{GB}$ w.r.t. the loss L and the Lebesgue measure has the following stochastic expansion :*

$$(7.2.1) \qquad n\hat{\theta}_{GB} = S + \frac{1}{3\sqrt{n}} Z_1 T^2 - \frac{I}{3n} S T^2 - \frac{b_3}{6b_2 n} T^2 + o_p\left(\frac{1}{n}\right),$$

where $S = n\left(\underline{\theta} + \overline{\theta}\right)/2$, $T = n\left(\overline{\theta} - \underline{\theta}\right)/2$ and $b_i = (d^i/du^i) L(0)$ $(i = 2, 3)$.

Proof. First we see that the generalized Bayes estimator is given by a solution $\hat{\theta}$ of the equation

$$(7.2.2) \qquad \int_{\underline{\theta}}^{\overline{\theta}} L^{(1)}\left(\hat{\theta} - \theta\right) \prod_{j=1}^{n} f_0(x_j - \theta) \, d\theta = 0,$$

where $L^{(1)}(u) = (d/du)L(u)$. Putting $t = \sqrt{n}\hat{\theta}$ and $u = \sqrt{n}\theta$, we have from (7.2.2)

$$(7.2.3) \qquad 0 = \int_{\sqrt{n}\underline{\theta}}^{\sqrt{n}\overline{\theta}} L^{(1)}\left(\frac{1}{\sqrt{n}}(t-u)\right) \exp\left\{\sum_{j=1}^{n} \log f_0\left(x_j - \frac{u}{\sqrt{n}}\right)\right\} \frac{du}{\sqrt{n}}.$$

Since $L^{(1)}(0) = 0$, it follows that

$$(7.2.4) \qquad L^{(1)}\left(\frac{1}{\sqrt{n}}(t-u)\right) = \frac{1}{\sqrt{n}}(t-u)L^{(2)}(0) + \frac{1}{2n}(t-u)^2 L^{(3)}(0) + \cdots$$

$$= \frac{b_2}{\sqrt{n}}(t-u) + \frac{b_3}{2n}(t-u)^2 + \cdots.$$

We also have

$$(7.2.5) \qquad \exp\left\{\sum_{j=1}^{n} \log f_0\left(x_j - \frac{u}{\sqrt{n}}\right)\right\}$$

$$= \exp\left[\sum_{j=1}^{n} \ell(x_j) - \frac{u}{\sqrt{n}}\sum_{j=1}^{n}\ell^{(1)}(x_j) + \frac{u^2}{2n}\sum_{j=1}^{n}\ell^{(2)}(x_j) + \cdots\right]$$

$$= \prod_{j=1}^{n} f_0(x_j)\left[1 - \frac{u}{\sqrt{n}}\sum_{j=1}^{n}\ell^{(1)}(x_j) + \frac{u^2}{2n}\sum_{j=1}^{n}\ell^{(2)}(x_j)\right.$$

$$\left. + \frac{u^2}{2n}\left\{\sum_{j=1}^{n}\ell^{(1)}(x_j)\right\}^2 + \cdots\right]$$

$$= \prod_{j=1}^{n} f_0(x_j)\left\{1 + uZ_1 + \frac{u^2}{2\sqrt{n}}(Z_2 - \sqrt{n}I) + \frac{u^2}{2}Z_1^2 + \cdots\right\}.$$

From (7.2.3) to (7.2.5) we obtain

$$(7.2.6)\ 0 = \int_{\sqrt{n}\underline{\theta}}^{\sqrt{n}\bar{\theta}}\left\{\frac{b_2}{\sqrt{n}}(t-u) + \frac{b_3}{2n}(t-u)^2\right\}\left\{1 + uZ_1 + \frac{u^2}{2}(Z_1^2 - I)\right\} du + o_p\left(\frac{1}{n\sqrt{n}}\right)$$

$$= \int_{\sqrt{n}\underline{\theta}}^{\sqrt{n}\bar{\theta}}\left\{b_2(t-u) + b_2 Z_1 u(t-u) + \frac{b_2}{2}(Z_1^2 - I)u^2(t-u) + \frac{b_3}{2\sqrt{n}}(t-u)^2\right.$$

$$\left. + \frac{b_3}{2\sqrt{n}}Z_1 u(t-u)^2 + \frac{b_3}{4\sqrt{n}}(Z_1^2 - I)u^2(t-u)^2\right\} du + o_p\left(\frac{1}{n\sqrt{n}}\right).$$

Since $S = n\left(\underline{\theta} + \bar{\theta}\right)/2$ and $T = n\left(\bar{\theta} - \underline{\theta}\right)/2$, it follows from (7.2.6) that

$$0 = 2b_2 T\left(t - \frac{S}{\sqrt{n}}\right) + 2b_2 Z_1 T\left\{\frac{t}{\sqrt{n}}S - \frac{1}{3n}(3S^2 + T^2)\right\}$$

$$+ \frac{b_2}{2}(Z_1^2 - I)\left\{\frac{2t}{3n}T(3S^2 + T^2) - \frac{2}{n\sqrt{n}}TS(S^2 + T^2)\right\}$$

$$+ \frac{b_3}{\sqrt{n}}T\left\{t^2 - \frac{2}{\sqrt{n}}tS + \frac{1}{3n}(3S^2 + T^2)\right\} + \frac{b_3 t^2}{6n\sqrt{n}}(Z_1^2 - I)T(3S^2 + T^2)$$

$$+ o_p\left(\frac{1}{n\sqrt{n}}\right),$$

which implies

$$b_2\sqrt{n}t\left\{1 + \frac{1}{\sqrt{n}}Z_1 S + \frac{b_3}{2b_2\sqrt{n}}t - \frac{b_3}{b_2 n}S + \frac{1}{6n}(Z_1^2 - I)(3S^2 + T^2)\right\}$$

$$= b_2 S + \frac{b_2}{3\sqrt{n}}Z_1(3S^2 + T^2) + \frac{b_2}{2n}(Z_1^2 - I)S(S^2 + T^2) - \frac{b_3}{6n}(3S^2 + T^2) + o_p\left(\frac{1}{n}\right).$$

Hence we have

$$
n\hat{\theta} = \sqrt{n}t = \left\{ S + \frac{1}{3\sqrt{n}}Z_1\left(3S^2 + T^2\right) + \frac{1}{2n}\left(Z_1^2 - I\right)S\left(S^2 + T^2\right) \right.
$$
$$
\left. - \frac{b_3}{6b_2 n}\left(3S^2 + T^2\right) \right\}\left\{ 1 - \frac{1}{\sqrt{n}}Z_1 S - \frac{b_3}{2b_2\sqrt{n}}t + \frac{b_3}{b_2 n}S \right.
$$
$$
\left. - \frac{1}{6n}\left(Z_1^2 - I\right)\left(3S^2 + T^2\right) + \frac{1}{n}Z_1^2 S^2 + \frac{b_3^2}{4b_2^2 n}t^2 \right\} + o_p\left(\frac{1}{n}\right)
$$
$$
= S + \frac{1}{3\sqrt{n}}Z_1 T^2 - \frac{I}{3n}ST^2 - \frac{b_3}{6b_2 n}T^2 + o_p\left(\frac{1}{n}\right).
$$

This completes the proof.

Theorem 7.2.2. *Under the conditions (A.7.1.1) to (A.7.1.3), the characteristic function of the generalized Bayes estimator $\hat{\theta}_{GB}$, up to the order n^{-1}, is given as follows:*

$$
\tilde{\phi}_n(t) = E\left[e^{2citn\left(\hat{\theta}_{GB} - \theta\right)} \right]
$$
$$
= \phi_0(t) + \frac{1}{n}\sum_{j=0}^{3}\alpha_j\phi_j(t) + \frac{1}{12c^2 n}\sum_{j=0}^{3}i\beta_j t\phi_j(t) - \frac{1}{18c^2 n}\sum_{j=0}^{4}\gamma_j t^2\phi_j(t) + o\left(\frac{1}{n}\right),
$$

where

$$
\alpha_0 = \frac{1}{12c^2}\left(h^- + 2I\right), \quad \alpha_1 = 1 + \frac{1}{12c^2}\left(5h^- - 2I\right), \quad \alpha_2 = -\frac{1}{2} - \frac{1}{4c^2}\left(2h^- - I\right),
$$
$$
\alpha_3 = \frac{1}{12c^2}\left(h^- - I\right), \quad \beta_0 = \beta_1 = -\left(h^+ + \frac{2b_3 c}{b_2}\right), \quad \beta_2 = -\left(2h^+ + \frac{b_3 c}{b_2}\right),
$$
$$
\beta_3 = h^+, \quad \gamma_0 = \gamma_1 = 6h^-, \quad \gamma_2 = 3h^-, \quad \gamma_3 = I, \quad \gamma_4 = \frac{I}{4},
$$
$$
\phi_0(t) = \frac{1}{1 + t^2}, \quad \phi_1(t) = \frac{1 - t^2}{(1 + t^2)^2}, \quad \phi_2(t) = \frac{2\left(1 - 3t^2\right)}{(1 + t^2)^3},
$$
$$
\phi_3(t) = \frac{6\left(1 - 6t^2 + t^4\right)}{(1 + t^2)^4}, \quad \phi_4(t) = \frac{24\left(1 - 10t^2 + 5t^4\right)}{(1 + t^2)^5}
$$

with $h^+ = f_0'(a + 0) + f_0'(b - 0)$ and $h^- = f_0'(a + 0) - f_0'(b - 0)$, and i denotes the imaginary unit.

Remark 7.2.1. Since for $j = 0, 1, 2, \ldots$,

$$
\phi_j(t) = \frac{1}{2}\int_{-\infty}^{\infty}|x|^j e^{-|x|}e^{itx}dx = \frac{j!}{2}\left\{ \frac{1}{(1 + it)^{j+1}} + \frac{1}{(1 - it)^{j+1}} \right\}
$$
$$
= \frac{j!}{(1 + t^2)^{j+1}}\left\{ 1 - \binom{j+1}{2}t^2 + \binom{j+1}{4}t^4 - \cdots \right\},
$$

it follows that

$$\frac{1}{2\pi} \int_{-\infty}^{\infty} \phi_j(t) e^{-itx} dt = \frac{1}{2} |x|^j e^{-|x|} \qquad (j = 0, 1, 2, \ldots).$$

Remark 7.2.2. As is seen Theorem 7.2.2, the characteristic function of $\hat{\theta}_{GB}$ depends on the sum h^+ and the difference h^- of unilateral differential coefficients $f_0'(a+0)$ and $f_0'(b-0)$ at the extreme points a and b, in addition to the Fisher information I. Hence the second order asymptotic behavior of $\hat{\theta}_{GB}$ is affected through h^+ and h^- at a and b, and through I in the inside of the interval (a, b).

Proof. From (7.2.1) it follows that the characteristic function $\Phi_n(t)$ of $n \left(\hat{\theta}_{GB} - \theta \right)$ is given by

$$(7.2.7) \qquad \Phi_n(t) = E_\theta \left[e^{itn(\hat{\theta}_{GB} - \theta)} \right]$$

$$= E_0 \left[e^{itS} \left\{ 1 + \frac{it}{3\sqrt{n}} E_0 \left(Z_1 T^2 | S \right) - \frac{itI}{3n} S E_0 \left(T^2 | S \right) \right. \right.$$

$$\left. \left. - \frac{ib_3 t}{6b_2 n} E_0 \left(T^2 | S \right) - \frac{t^2}{18n} E_0 \left(Z_1^2 T^4 | S \right) \right\} \right] + o \left(\frac{1}{n} \right),$$

where $E_0(\cdot | S)$ denotes the conditional expectation given S. Since

$$E_0 \left(Z_1 | S, T \right) = -\frac{h^+}{c\sqrt{n}} + \frac{1}{\sqrt{n}} \left(h^- S + h^+ T \right) + o_p \left(\frac{1}{n} \right),$$

it follows that

$$(7.2.8) E_0 \left(Z_1 T^2 | S \right) = E_0 \left[E_0 \left(Z_1 | S, T \right) T^2 | S \right]$$

$$= -\frac{h^+}{c\sqrt{n}} E_0 \left(T^2 | S \right) + \frac{h^-}{\sqrt{n}} S E_0 \left(T^2 | S \right) + \frac{h^+}{\sqrt{n}} E_0 \left(T^3 | S \right) + o_p \left(\frac{1}{n} \right),$$

where $h^+ = f_0'(a+0) + f_0'(b-0)$ and $h^- = f_0'(a+0) - f_0'(b-0)$. Since

$$E_0 \left(Z_1^2 | S, T \right) = I + O_p \left(\frac{1}{n} \right),$$

it follows that

$$(7.2.9) \qquad E_0 \left(Z_1^2 T^4 | S \right) = E_0 \left[E_0 \left(Z_1^2 T^4 | S, T \right) | S \right]$$

$$= E_0 \left[T^4 E_0 \left(Z_1^2 | S, T \right) | S \right]$$

$$= I E_0 \left(T^4 | S \right) + O_p \left(\frac{1}{n} \right).$$

Since the conditional density $g(T|S)$ of T, given S, is obtained by

$$g(T|S) = \begin{cases} 2ce^{-2c(T-|S|)} & \text{for} \quad T > |S|; \\ 0 & \text{for} \quad T \le |S|, \end{cases}$$

we have

$$E_0(T|S) = |S| + \frac{1}{2c}; \quad E_0\left(T^2|S\right) = S^2 + \frac{1}{c}|S| + \frac{1}{2c^2};$$

$$E_0\left(T^3|S\right) = |S|^3 + \frac{3}{2c}S^2 + \frac{3}{2c^2}|S| + \frac{3}{4c^3};$$

$$E_0\left(T^4|S\right) = S^4 + \frac{2}{c}|S|^3 + \frac{3}{c^2}S^2 + \frac{3}{c^3}|S| + \frac{3}{2c^4}.$$

From (7.2.8) and (7.2.9) we obtain

$$E_0\left(Z_1 T^2|S\right) = -\frac{h^+}{c\sqrt{n}}\left(S^2 + \frac{1}{c}|S| + \frac{1}{2c^2}\right) + \frac{h^-}{\sqrt{n}}S\left(S^2 + \frac{1}{c}|S| + \frac{1}{2c^2}\right)$$

$$+ \frac{h^+}{\sqrt{n}}\left(|S|^3 + \frac{3}{2c}S^2 + \frac{3}{2c^2}|S| + \frac{3}{4c^3}\right);$$

$$E_0\left(Z_1^2 T^4|S\right) = I\left(S^4 + \frac{2}{c}|S|^3 + \frac{3}{c^2}S^2 + \frac{3}{c^3}|S| + \frac{3}{2c^4}\right),$$

hence by (7.2.7)

$$\Phi_n(t) = E_0\left[e^{itS}\left\{1 + \frac{it}{3n}\left((h^- - I)S^3 + h^+|S|^3 + \left(\frac{h^+}{2c} - \frac{b_3}{2b_2}\right)S^2 + \frac{1}{c}(h^- - I)S|S|\right.\right.\right.$$

$$+ \frac{1}{2c^2}(h^- - I)S + \left(\frac{h^+}{2c^2} - \frac{b_3}{2b_2c}\right)|S| + \frac{h^+}{4c^3} - \frac{b_3}{4b_2c^2}\right)$$

$$\left.\left.\left. - \frac{It^2}{18n}\left(S^4 + \frac{2}{c}|S|^3 + \frac{3}{c^2}S^2 + \frac{3}{c^3}|S| + \frac{3}{2c^4}\right)\right\}\right] + o\left(\frac{1}{n}\right).$$

By Remark 7.2.1 we have

$$\tilde{\Phi}_n(t) = \Phi_n(2ct) = \phi_0(t) + \frac{1}{n}\left[\frac{1}{12c^2}(h^- + 2I)\phi_0(t) + \left\{1 + \frac{1}{12c^2}(5h^- - 2I)\right\}\phi_1(t)\right.$$

$$\left. - \left\{\frac{1}{2} + \frac{1}{4c^2}(2h^- - I)\right\}\phi_2(t) + \frac{1}{12c^2}(h^- - I)\phi_3(t)\right]$$

$$+ \frac{1}{12c^2 n}\left\{-\left(h^+ + \frac{2b_3c}{b_2}\right)it\phi_0(t) - \left(h^+ + \frac{2b_3c}{b_2}\right)it\phi_1(t)\right.$$

$$\left. - \left(2h^+ + \frac{b_3c}{b_2}\right)it\phi_2(t) + h^+ it\phi_3(t)\right\} - \frac{1}{18c^2 n}\left\{6h^- t^2\phi_0(t)\right.$$

$$\left. + 6h^- t^2\phi_1(t) + 3h^- t^2\phi_2(t) + It^2\phi_3(t) + \frac{I}{4}t^2\phi_4(t)\right\} + o\left(\frac{1}{n}\right)$$

$$= \phi_0(t) + \frac{1}{n}\sum_{j=0}^{3}\alpha_j\phi_j(t) + \frac{1}{12c^2 n}\sum_{j=0}^{3} i\beta_j t\phi_j(t)$$

$$- \frac{1}{18c^2 n}\sum_{j=0}^{4}\gamma_j t^2\phi_j(t) + o\left(\frac{1}{n}\right) \qquad \text{(say)}.$$

This completes the proof.

Corollary 7.2.1. *Under the conditions (A.7.1.1) to (A.7.1.3), the asymptotic density of $2cn\left(\hat{\theta}_{GB} - \theta\right)$, up to the order n^{-1}, is given by*

$$\tilde{g}_n(x) = g_{00}(x) + \frac{1}{n}\sum_{j=0}^{3}\alpha_j g_{0j}(x) + \frac{1}{12c^2 n}\sum_{j=0}^{3}\beta_j g_{1j}(x) - \frac{1}{18c^2 n}\sum_{j=0}^{4}\gamma_j g_{2j}(x) + o\left(\frac{1}{n}\right),$$

where α_j's, β_j's and γ_j's are given in Theorem 7.2.2, and

$$g_{0j}(x) = \frac{1}{2}|x|^j e^{-|x|} \quad (j = 0, 1, 2, 3),$$

$$g_{10}(x) = \frac{1}{2}(\text{sgn}x)e^{-|x|}, \quad g_{11}(x) = \frac{1}{2}(x - \text{sgn}x)e^{-|x|},$$

$$g_{12}(x) = \left(\frac{x^2}{2}\text{sgn}x - x\right)e^{-|x|}, \quad g_{13}(x) = \frac{x^2}{2}(x - 3\text{sgn}x)e^{-|x|},$$

$$g_{20}(x) = -\frac{1}{2}e^{-|x|}, \quad g_{21}(x) = -\left(\frac{|x|}{2} - 1\right)e^{-|x|},$$

$$g_{22}(x) = -\left(1 - 2|x| + \frac{x^2}{2}\right)e^{-|x|}, \quad g_{23}(x) = -\left(\frac{|x|^3}{2} - 3x^2 + 3|x|\right)e^{-|x|},$$

$$g_{24}(x) = -\left(\frac{x^4}{2} - 4|x|^3 + 6x^2\right)e^{-|x|}.$$

The proof follows from Theorem 7.2.2, Remark 7.2.1 and Fourier inverse transforms.

Corollary 7.2.2. *Under the conditions (A.7.1.1) to (A.7.1.3), the asymptotic density of $n\left(\hat{\theta}_{GB} - \theta\right)$, up to the order n^{-1}, is given by*

$$g_n(x) = ce^{-2c|x|}\left[1 + \frac{1}{n}\sum_{j=0}^{3}(2c)^j\alpha_j|x|^j + \frac{1}{12c^2 n}\left\{6ch^+x + 8c^3h^+x^3\right.\right.$$

$$- \left(5h^+ + \frac{b_3 c}{b_2}\right)\text{sgn}x\right\} - \frac{1}{9c^2 n}\left\{6c\left(h^- - I\right)|x| - 6c^2\left(h^- - I\right)x^2\right.$$

$$\left.\left. + 4c^3 I|x|^3 - 2c^4 Ix^4\right\}\right] + o\left(\frac{1}{n}\right),$$

where α_j's are given in Theorem 7.2.2.

The proof is an immediate consequence of Corollary 7.2.1.

Corollary 7.2.3. *Assume that the conditions (A.7.1.1) to (A.7.1.3) hold. If $b_3 = 0$ and $h^+ = 0$, then the asymptotic density of $n\left(\hat{\theta}_{GB} - \theta\right)$, up to the order n^{-1}, is given by*

$$g_n(x) = ce^{-2c|x|}\left\{1 + \frac{1}{n}\left(k_0 + k_1|x| + k_2 x^2 + k_3|x|^3 + k_4 x^4\right)\right\} + o\left(\frac{1}{n}\right),$$

where

$$k_0 = \frac{1}{12c^2}\left(h^- + 2I\right), \quad k_1 = 2c + \frac{1}{6c}\left(h^- + 2I\right), \quad k_2 = -2c^2 - \frac{1}{3}\left(4h^- - I\right),$$

$$k_3 = \frac{2}{9}c\left(3h^- - 5I\right), \quad k_4 = \frac{2}{9}c^2 I.$$

The proof is straightforward from Corollary 7.2.2.

Remark 7.2.3. The condition of $h^+ = 0$ is satisfied if $f_0(x)$ is symmetric around $x = (a+b)/2$. Such cases will be discussed in the next section.

Theorem 7.2.3. *Under the same conditions as in Corollary 7.2.3, it holds that for the generalized Bayes estimator $\hat{\theta}_{GB}$*

$$P_\theta^n\left\{n\left|\hat{\theta}_{GB} - \theta\right| \leq t\right\}$$

$$= 1 - e^{-2ct} + \frac{1}{n}e^{-2ct}\left[\frac{1}{6c}\left(h^- + 2I\right)t + \left\{2c^2 + \frac{1}{3}\left(h^- + 2I\right)\right\}t^2\right.$$

$$\left. - \frac{2}{3}c\left(h^- - I\right)t^3 - \frac{2}{9}c^2 I t^4\right] + o\left(\frac{1}{n}\right)$$

for all $t > 0$.

Proof. This results from an immediate calculation of the integral of $g(x)$ given in Corollary 7.2.3.

7.3. SECOND ORDER ASYMPTOTIC BOUND IN SYMMETRICALLY TRUNCATED DENSITIES

We assume that $f_0(x)$ is symmetric around $x = (a+b)/2$. Let θ_0 be any fixed element of Θ and denote by $\mathbf{A}_2$ a class of the all second order AMU estimators of θ. In a similar way to Akahira (1982) and Akahira and Takeuchi (1981), we shall obtain the upper bound for the asymptotic concentration probability (ACP) for $\hat{\theta}_n(\in \mathbf{A}_2)$ around θ_0, i.e., $P_{\theta_0}^n\left\{n\left|\hat{\theta}_n - \theta_0\right| \leq t\right\}$ in the class $\mathbf{A}_2$, up to the order n^{-1}. In order to do so, it is enough to get the upper bound for

$$(7.3.1) \qquad\qquad P_{\theta_0 - tn^{-1}}^n\left\{\hat{\theta}_n \leq \theta_0\right\} - P_{\theta_0 + tn^{-1}}^n\left\{\hat{\theta}_n \leq \theta_0\right\}$$

in the class $\mathbf{A}_2$, up to the order n^{-1}, since $\hat{\theta}_n$ is second order AMU.

In a similar way to the fundamental lemma of Neyman and Pearson, it is shown that $\hat{\theta}_n^*$ maximizing (7.3.1) is given by

(7.3.2) $\quad \phi_n^*(\tilde{x}_n) = \begin{cases} 1, & \displaystyle\prod_{i=1}^{n} f_0\left(x_i - \theta_0 + tn^{-1}\right) > \prod_{i=1}^{n} f_0\left(x_i - \theta_0 - tn^{-1}\right); \\[4mm] 0, & \displaystyle\prod_{i=1}^{n} f_0\left(x_i - \theta_0 + tn^{-1}\right) < \prod_{i=1}^{n} f_0\left(x_i - \theta_0 - tn^{-1}\right), \end{cases}$

where $\tilde{x}_n = (x_1, \ldots, x_n)$. Further we assume that $\lim_{x \to b-0} f_0'(x) = -\lim_{x \to a+0} f_0'(x) = h$. Using (7.3.2) we have the following result.

Theorem 7.3.1. *Assume that the conditions (A.7.1.1) to (A.7.1.3) hold. Then for any $\hat{\theta}_n \in \mathbf{A}_2$, any $\theta \in \Theta$ and any $t > 0$*

$$P_\theta^n \left\{ n \left| \hat{\theta}_n - \theta \right| \leq t \right\} \leq 1 - e^{-2ct} + \sqrt{\frac{2I}{\pi n}} t e^{-2ct} + \frac{2}{n}\left(c^2 - h\right) t^2 e^{-2ct} + o\left(\frac{1}{n}\right).$$

Remark 7.3.1. The upper bound for the ACP for $\hat{\theta}_n \in \mathbf{A}_2$ is given up to the order n^{-1} in the above theorem. Then it is seen that the first term of the upper bound is $1 - e^{-2ct}$ and the next term is of order $n^{-1/2}$, i.e. the 3/2-th order, since $c_n = n$ and $c_n^{k-1} = n^{1/2}$ for $k = 3/2$, but not of order n^{-1} i.e., not the second order. It is noted that the term of the order $n^{-1/2}$ in the upper bound is positive for $I > 0$ and all $t > 0$. On the other hand, as is seen from Theorems 7.2.3 and 7.3.1, the generalized Bayes estimator $\hat{\theta}_{GB}$ is first order two-sided asymptotically efficient, but there exist no term of the order $n^{-1/2}$ in the ACP for $\hat{\theta}_{GB}$. Hence, if $I > 0$, $\hat{\theta}_{GB}$ is not 3/2-th order two sided asymptotically efficient. It is remarked that $I = 0$ if and only if $f(x)$ is an uniform density on the interval (a, b). From Theorem 7.3.1 it is also noted that the second order upper bound is affected by the endpoints a and b of the support (a, b) of the density $f_0(x)$ through $c^2 - h = f_0^2(a+0) + f_0'(a+0) = f_0^2(b-0) - f_0'(b-0)$ and the inner points of the interval through I.

Proof of Theorem 7.3.1. We put

$A = \left\{ \tilde{x}_n : \underline{\theta} < \theta_0 - tn^{-1}, \overline{\theta} < \theta_0 + tn^{-1} \right\}$, $B = \left\{ \tilde{x}_n : \underline{\theta} > \theta_0 - tn^{-1}, \overline{\theta} > \theta_0 + tn^{-1} \right\}$,

$C = \left\{ \tilde{x}_n : \underline{\theta} < \theta_0 - tn^{-1}, \overline{\theta} > \theta_0 + tn^{-1} \right\}$, $D = \left\{ \tilde{x}_n : \dfrac{1}{\sqrt{n}} Z_1 \leq \theta_0 \right\}$,

$D' = \left\{ \tilde{x}_n : \dfrac{1}{\sqrt{n}} Z_1 > \theta_0 \right\}$.

From (7.3.2) we have

$$\phi_n^*(\tilde{x}_n) = \begin{cases} 1 & \text{for} \quad \tilde{x}_n \in A \cup (C \cap D); \\ 0 & \text{for} \quad \tilde{x}_n \in B \cup (C \cap D'). \end{cases}$$

Then we have for any $\hat{\theta}_n \in \mathbf{A}_2$

$$(7.3.3) \quad P_{\theta_0 - tn^{-1}}^n \left\{ \hat{\theta}_n \le \theta_0 \right\} - P_{\theta_0 + tn^{-1}}^n \left\{ \hat{\theta}_n \le \theta_0 \right\}$$

$$\le E_{\theta_0 - tn^{-1}}^n (\phi_n^*) - E_{\theta_0 + tn^{-1}}^n (\phi_n^*)$$

$$= P_{\theta_0 - tn^{-1}}^n (A) - P_{\theta_0 + tn^{-1}}^n (A) + P_{\theta_0 - tn^{-1}}^n (C \cap D) - P_{\theta_0 + tn^{-1}}^n (C \cap D).$$

Putting $U = n \left(\overline{\theta} - \theta_0 \right)$ and $V = n \left(\underline{\theta} - \theta_0 \right)$, we have, as the joint density of (U, V),

$$g_n(u,v) = \begin{cases} \dfrac{n-1}{n} \left\{ F \left(b + \dfrac{v}{n} \right) - F \left(a + \dfrac{u}{n} \right) \right\}^{n-2} f \left(a + \dfrac{u}{n} \right) f \left(b + \dfrac{v}{n} \right) \\ \qquad\qquad\qquad\qquad\qquad\qquad\qquad\qquad \text{for} \quad u > 0, v < 0, \\ 0 \qquad\qquad\qquad\qquad\qquad\qquad\qquad\qquad\quad \text{otherwise}, \end{cases}$$

where $F(x) = \int_{-\infty}^x f(t)dt$, hence, by its expansion,

$$g_n(u,v) = \begin{cases} c^2 e^{-c(u-v)} \left[1 + \dfrac{1}{n} \left\{ -1 + 2c(u-v) + \dfrac{h}{4} \left((u+v)^2 + (u-v)^2 \right) \right. \right. \\ \qquad\qquad \left. \left. -\dfrac{c^2}{2}(u-v)^2 - \dfrac{h}{c}(u-v) \right\} \right] + o \left(\dfrac{1}{n} \right) \quad \text{for} \quad u > 0, v < 0, \\ 0 \qquad\qquad\qquad\qquad\qquad\qquad\qquad\qquad\qquad \text{otherwise}. \end{cases}$$

Since, from the expansion of $g_n(u,v)$, the asymptotic density of (S, T), up to the order $o(1/n)$, is given by

$$(7.3.4)$$

$$f_n(S,T) = \begin{cases} 2c^2 e^{-2cT} \left[1 + \dfrac{1}{n} \left\{ -1 + 4cT + h \left(T^2 + S^2 \right) - 2c^2 T^2 \right. \right. \\ \qquad\qquad\qquad \left. \left. -\dfrac{2h}{c}T \right\} \right] + o \left(\dfrac{1}{n} \right) \quad \text{for} \quad -T < S < T, \, 0 < T, \\ 0 \qquad\qquad\qquad\qquad\qquad\qquad\qquad\qquad \text{otherwise}, \end{cases}$$

it follows that

$$(7.3.5) \quad P_{\theta_0 - tn^{-1}}^n (A) = P_{\theta_0 - tn^{-1}}^n \left\{ \underline{\theta} < \theta_0 - tn^{-1}, \overline{\theta} < \theta_0 + tn^{-1} \right\}$$

$$= P_{\theta_0 - tn^{-1}}^n \left\{ n \left(\underline{\theta} - (\theta_0 - tn^{-1}) \right) < 0, n \left(\overline{\theta} - (\theta_0 - tn^{-1}) \right) < 2t \right\}$$

$$= P_0^n \left\{ n\underline{\theta} < 0, n\overline{\theta} < 2t \right\}$$

$$= P_0^n \{ S - T < 0, S + T < 2t \}$$

$$= \left(\int_0^t \int_S^{-S+2t} + \int_{-\infty}^0 \int_{-S}^{-S+2t} \right) f_n(T,S) dT dS$$

$$= 1 - e^{-2ct} + \dfrac{2}{n} \left(c^2 - h \right) t^2 e^{-2ct} + o \left(\dfrac{1}{n} \right).$$

Similarly we have

(7.3.6) $$P^n_{\theta_0 + tn^{-1}}(A) = o\left(\frac{1}{n}\right).$$

Then it follows that the conditional probability of the event $\{Z_1/\sqrt{n} \le \theta_0\}$ given S and T on the set C, up to the order n^{-1}, is given by

$$P^n_{\theta_0 \mp tn^{-1}}\left\{\frac{Z_1}{\sqrt{n}} \le \theta_0 \,\Big|\, S, T\right\} = \frac{1}{2} + \frac{2hS \pm It}{\sqrt{2\pi In}} + o\left(\frac{1}{n}\right),$$

hence, by (7.3.4),

$$(7.3.7)\quad P^n_{\theta_0 \mp tn^{-1}}(C \cap D) = \int_C P^n_{\theta_0 \mp tn^{-1}}\left\{\frac{Z_1}{\sqrt{n}} \le \theta_0 \,\Big|\, S, T\right\} f_n(S, T)\,dS\,dT$$

$$= \frac{1}{2}e^{-2ct} \pm \frac{1}{2}\sqrt{\frac{2I}{\pi n}}te^{-2ct}$$

$$+ \frac{1}{n}\left\{\left(c + \frac{h}{2c}\right)t - \left(c^2 - \frac{h}{2}\right)t^2\right\}e^{-2ct} + o\left(\frac{1}{n}\right),$$

where the signs $+$ and $-$ should be read consistently. From (7.3.3) to (7.3.7) we have

$$P^n_{\theta_0}\left\{n\left|\hat\theta_n - \theta_0\right| \le t\right\} \le 1 - e^{-2ct} + \sqrt{\frac{2I}{\pi n}}te^{-2ct} + \frac{2}{n}\left(c^2 - h\right)t^2 e^{-2ct} + o\left(\frac{1}{n}\right).$$

Since θ_0 is arbitrary, the conclusion of the theorem follows.

7.4. MAXIMUM PROBABILITY ESTIMATION

In this section, following Akahira (1991a), we obtain the ACP of the maximum probability estimator (MPE) up to the order $o\left(n^{-1/2}\right)$ and show that the MPE attains the upper bound for the ACP of $\hat\theta_n(\in \mathbf{A}_{3/2})$ up to the order $o\left(n^{-1/2}\right)$ at any specified point only when $h = 0$.

Let r be any fixed positive number. The MPE $\hat\theta^r_{MP}$ of θ is defined as that value of d maximizing

$$\int_{d-(r/n)}^{d+(r/n)} \prod_{i=1}^n f_0(x_i - \theta)\,d\theta$$

(Weiss and Wolfowitz (1967, 1974)). Note that if we consider a gain function as

$$G_r(\theta - d) = \begin{cases} 1 & \text{for } |\theta - d| < r/n, \\ 0 & \text{for } |\theta - d| \ge r/n, \end{cases}$$

then the MPE $\hat\theta^r_{MP}$ may be essentially regarded as a generalized Bayes estimator. In addition to the conditions (A.7.1.1) to (A.7.1.3) we assume that $\lim_{x \to b-0} f'_0(x) =$

$-\lim_{x\to a+0} f_0'(x) = h \leq 0$, $I > 0$, and $f_0(x)$ is symmetric around $x = (a + b)/2$. Then the MPE $\hat{\theta}_{MP}^r$ is shown to be

$$(7.4.1) \qquad \hat{\theta}_{MP}^r = \begin{cases} S/n & \text{for } T \leq r, \\ \overline{\theta} - (r/n) & \text{for } T > r,\ \hat{\theta}_0 \geq \overline{\theta} - (r/n), \\ \underline{\theta} + (r/n) & \text{for } T > r,\ \hat{\theta}_0 \leq \underline{\theta} + (r/n), \\ \hat{\theta}_0 & \text{for } \underline{\theta} + (r/n) < \hat{\theta}_0 < \overline{\theta} - (r/n), \end{cases}$$

where $\hat{\theta}_0$ is the maximum likelihood estimator whose local uniqueness in the interval $(\underline{\theta}, \overline{\theta})$ is guaranteed since $E_\theta\left[\{(\partial^2/\partial\theta^2)\log f(X-\theta)\}\,\chi_{(a,b)}(X-\theta)\right] < 0$, where $\chi_{(a,b)}(\cdot)$ is an indicator of the interval (a, b). Since $f_0(x)$ is symmetric, it is seen that the $\hat{\theta}_{MP}^r$ belongs to the class $\mathbf{A}_{3/2}$. Next we obtain the ACP of the MPE $\hat{\theta}_{MP}^r$ up to the order $o\left(n^{-1}\right)$.

Theorem 7.4.1. *Assume that the conditions (A.7.1.1) to (A.7.1.3) and the above hold. Then the ACP of the MPE $\hat{\theta}_{MP}^r$ is given by*

$$P_\theta^n\left\{n\left|\hat{\theta}_{MP}^r - \theta\right| \leq t\right\}$$
$$= \begin{cases} 1 - e^{-2ct} + \dfrac{4ht}{\sqrt{2\pi In}}e^{-2cr} + \sqrt{\dfrac{2I}{\pi n}}\left\{t - \dfrac{c}{2}(t-r)^2\right\}e^{-2cr} + o\left(\dfrac{1}{\sqrt{n}}\right) & \text{for } t \leq r, \\ 1 - e^{-c(r+t)} + \dfrac{2h}{\sqrt{2\pi In}}(r+t)e^{-c(r+t)} + \sqrt{\dfrac{2I}{\pi n}}te^{-c(r+t)} + o\left(\dfrac{1}{\sqrt{n}}\right) & \text{for } t > r. \end{cases}$$

Further, if $h = 0$, then the MPE $\hat{\theta}_{MP}^r$ is 3/2-th order two-sided asymptotically efficient at $t = r$ in the class $\mathbf{A}_{3/2}$ in the sense that the ACP of the MPE $\hat{\theta}_{MP}^r$ attains the upper bound given by Theorem 7.3.1 at $t = r$ up to the order $o\left(n^{-1/2}\right)$ in the class $\mathbf{A}_{3/2}$.

Remark 7.4.1. As is seen from Theorems 7.3.1 and 7.4.1, the MPE $\hat{\theta}_{MP}^r$ is not first order two-sided asymptotically efficient, but it has the most ACP at the point r up to the order $o\left(n^{-1/2}\right)$ in the class $\mathbf{A}_{3/2}$ only when $h = 0$. If $h < 0$, then it follows from Theorems 7.3.1 and 7.4.1 that the MPE $\hat{\theta}_{MP}^r$ does not attain the bound up to the order $o\left(n^{-1/2}\right)$ at $t = r$.

Remark 7.4.2. It is noted that the order c_n of consistency is equal to n, but there exists a term of order $n^{-1/2}$ in the ACP of the MPE. Hence, in the non-regular case, the order k of asymptotic efficiency is equal to a fraction 3/2. This is quite different from the fact that the order k takes only positive integer in regular cases.

Proof. Without loss of generality, we assume that the parameter θ is equal to 0. From (7.4.1) it follows that

$$(7.4.2) \qquad P_0\left\{n\left|\hat{\theta}_{MP}^r\right| \leq t\right\}$$
$$= P_0\left\{|S| \leq t, T \leq r\right\} + P_0\left\{\left|n\overline{\theta} - r\right| \leq t, T > r, n\hat{\theta}_0 \geq n\overline{\theta} - r\right\}$$

$$+ P_0 \left\{ |n\underline{\theta} + r| \le t, T > r, n\hat{\theta}_0 \le n\underline{\theta} + r \right\}$$

$$+ P_0 \left\{ n|\hat{\theta}_0| \le t, n\underline{\theta} + r < n\hat{\theta}_0 < n\overline{\theta} - r \right\}$$

$$= p_1 + p_2 + p_3 + p_4 \qquad \text{(say)}.$$

(i) p_1: Since the asymptotic density of (S, T) is given by

$$f_n(S, T) = \begin{cases} 2c^2 e^{-2cT} + O\left(\dfrac{1}{n}\right) & \text{for} \quad -T < S < T, \ T > 0, \\ 0 & \text{otherwise}, \end{cases}$$

it follows that for $t \le r$

$$p_1 = \int_0^t \int_{-T}^T 2c^2 e^{-2cT} dS dT + \int_t^r \int_{-t}^t 2c^2 e^{-2cT} dS dT + O\left(\frac{1}{n}\right)$$

$$= -2cte^{-2cr} - e^{-2ct} + 1 + O\left(\frac{1}{n}\right),$$

and for $t > r$

$$p_1 = \int_0^r \int_{-T}^T 2c^2 e^{-2cT} dS dT + O\left(\frac{1}{n}\right) = -2cre^{-2cr} - e^{-2cr} + 1 + O\left(\frac{1}{n}\right).$$

(ii) p_2: Putting $u = n\overline{\theta}$ and $v = n\underline{\theta}$, we have the asymptotic density of (u, v) as

$$g_n(u, v) = \begin{cases} c^2 e^{-c(u-v)} + O\left(\dfrac{1}{n}\right) & \text{for} \quad u > 0, v < 0, \\ 0 & \text{otherwise}. \end{cases}$$

We also put $Z_1 = -(1/\sqrt{n}) \sum_{i=1}^n \ell^{(1)}(X_i)$, where $\ell^{(1)} = (d/dx)\ell(x)$ with $\ell(x) = \log f_0(x)$ for $a < x < b$. Then the asymptotic conditional cumulants of Z_1 given $\underline{\theta}$ and $\overline{\theta}$ on the set $\{\underline{\theta}, \overline{\theta} | u - v > 2r\}$, up to the third, are given by

$$E_0\left(Z_1 | \underline{\theta}, \overline{\theta}\right) = -\frac{h\left(n\underline{\theta} + n\overline{\theta}\right)}{\sqrt{n}} + O_p\left(\frac{1}{n\sqrt{n}}\right),$$

$$V_0\left(Z_1 | \underline{\theta}, \overline{\theta}\right) = I + O_p\left(\frac{1}{n}\right),$$

$$K_{3,0}\left(Z_1 | \underline{\theta}, \overline{\theta}\right) = O_p\left(\frac{1}{n}\right),$$

hence the Edgeworth expansion of the conditional distribution of Z_1 given $\underline{\theta}$ and $\overline{\theta}$ on the set is obtained by

$$(7.4.3) \qquad P_0\left\{Z_1 \le t | \underline{\theta}, \overline{\theta}\right\} = \Phi\left(\frac{t}{\sqrt{I}}\right) + \frac{h(u+v)}{\sqrt{In}} \phi\left(\frac{t}{\sqrt{I}}\right) + o\left(\frac{1}{\sqrt{n}}\right),$$

where $\Phi(x) = \int_{-\infty}^{x} \phi(t)dt$ with $\phi(t) = (1/\sqrt{2\pi})\,e^{-t^2/2}$. From (7.4.3) it follows that the conditional probability of the event $\{n\hat{\theta}_0 \geq u - r\}$ given $\underline{\theta}$ and $\bar{\theta}$ on the set $\{\underline{\theta}, \bar{\theta} \,|\, u - v > 2r\}$ is given by

$$P_0\left\{n\hat{\theta}_0 \geq u - r \,\Big|\, \underline{\theta}, \bar{\theta}\right\} = P_0\left\{Z_1 \geq \frac{I(u-r)}{\sqrt{n}} \,\Big|\, \underline{\theta}, \bar{\theta}\right\}$$

$$= 1 - \Phi\left(\sqrt{\frac{I}{n}}(u-r)\right) - \frac{h(u+v)}{\sqrt{In}}\phi\left(\sqrt{\frac{I}{n}}(u-r)\right)$$

$$+ o\left(\frac{1}{\sqrt{n}}\right)$$

$$= \frac{1}{2} - \sqrt{\frac{I}{2\pi n}}(u-r) - \frac{h(u+v)}{\sqrt{2\pi In}} + o\left(\frac{1}{\sqrt{n}}\right).$$

Hence we have for $t \leq r$

$$p_2 = P_0\left\{|u - r| \leq t, \frac{(u-v)}{2} > r, n\hat{\theta}_0 \geq u - r\right\}$$

$$= \int_{r-t}^{r+t} \int_{-\infty}^{u-2r} c^2 e^{-c(u-v)}\left\{\frac{1}{2} - \sqrt{\frac{I}{2\pi n}}(u-r) - \frac{h(u+v)}{\sqrt{2\pi In}}\right\} dv\,du + o\left(\frac{1}{\sqrt{n}}\right)$$

$$= cte^{-2cr} + \frac{2ht}{\sqrt{2\pi In}}e^{-2cr} + o\left(\frac{1}{\sqrt{n}}\right),$$

and also for $t > r$

$$p_2 = \left(\int_0^{2r} \int_{-\infty}^{u-2r} + \int_{2r}^{r+t} \int_{-\infty}^0\right) c^2 e^{-c(u-v)}\left\{\frac{1}{2} - \sqrt{\frac{I}{2\pi n}}(u-r) - \frac{h(u+v)}{\sqrt{2\pi In}}\right\} dv\,du$$

$$+ o\left(\frac{1}{\sqrt{n}}\right)$$

$$= cre^{-2cr} + \frac{2hre^{-2cr}}{\sqrt{2\pi In}} + \left\{\frac{1}{2} + \sqrt{\frac{I}{2\pi n}}\left(r - \frac{1}{c}\right)\right\}\left(e^{-2cr} - e^{-c(r+t)}\right)$$

$$+ \left(\sqrt{\frac{I}{2\pi n}} + \frac{h}{\sqrt{2\pi In}}\right)\left\{(r+t)e^{-c(r+t)} - 2re^{-2cr}\right\} + o\left(\frac{1}{\sqrt{n}}\right).$$

(iii) p_3: From the symmetry of $f_0(x)$ and (7.4.1) it follows that

$$p_3 = P_0\left\{|n\underline{\theta} + r| \leq t, T > r, n\hat{\theta}_0 \leq n\underline{\theta} + r\right\}$$

$$= P_0\left\{|n\bar{\theta} - r| \leq t, T > r, n\hat{\theta}_0 \geq n\bar{\theta} - r\right\} = p_2,$$

whose value is obtained in case (ii).

(iv) p_4: In this case we have

(7.4.4)
$$p_4 = P_0 \left\{ -t \leq n\hat{\theta}_0 \leq t, \ v+r < n\hat{\theta}_0 < u-r \right\}$$

$$= P_0 \left\{ \max(-t, v+r) \leq n\hat{\theta}_0 \leq \min(t, u-r) \right\}$$

$$= \int\int_{\substack{v<u-2r \\ v<0<u}} P_0 \left\{ \max(-t, v+r) \leq \frac{\sqrt{n}Z_1}{I} \leq \min(t, u-r) \middle| \underline{\theta}, \ \overline{\theta} \right\}$$

$$\cdot c^2 e^{-c(u-v)} du\,dv + O\left(\frac{1}{n}\right)$$

$$= \int\int_{\substack{v<u-2r \\ v<0<u}} P_0 \left\{ \frac{I}{\sqrt{n}} \max(-t, v+r) \leq Z_1 \leq \frac{I}{\sqrt{n}} \min(t, u-r) \middle| \underline{\theta}, \ \overline{\theta} \right\}$$

$$\cdot c^2 e^{-c(u-v)} du\,dv + O\left(\frac{1}{n}\right)$$

$$= \int\int_{\substack{v<u-2r \\ v<0<u}} \left[\sqrt{\frac{I}{2\pi n}} \{\min(t, u-r) - \max(-t, v+r)\} \right]$$

$$\cdot c^2 e^{-c(u-v)} du\,dv + O\left(\frac{1}{n}\right).$$

Then we obtain for $t \leq r$

$$\int\int_{\substack{v<u-2r \\ v<0<u}} \{\min(t, u-r)\} c^2 e^{-c(u-v)} du\,dv$$

$$= \int_0^{t+r} \int_{-\infty}^{u-2r} (u-r) c^2 e^{-c(u-v)} dv\,du$$

$$+ \left(\int_{t+r}^{2r} \int_{-\infty}^{u-2r} + \int_{2r}^{\infty} \int_{-\infty}^{0} \right) t c^2 e^{-c(u-v)} dv\,du$$

$$= \left\{ t - \frac{c(t-r)^2}{2} \right\} e^{-2cr},$$

and for $t > r$

$$\int\int_{\substack{v<u-2r \\ v<0<u}} \{\min(t, u-r)\} c^2 e^{-c(u-v)} du\,dv$$

$$= \left(\int_0^{2r} \int_{-\infty}^{u-2r} + \int_{2r}^{t+r} \int_{-\infty}^{0} \right) (u-r) c^2 e^{-c(u-v)} dv\,du + \int_{t+r}^{\infty} \int_{-\infty}^{0} t c^2 e^{-c(u-v)} dv\,du$$

$$= re^{-2cr} - \frac{1}{c} \left(e^{-c(t+r)} - e^{-2cr} \right).$$

We also have for $t \leq r$

$$\iint\limits_{\substack{v<u-2r \\ v<0<u}} \{\max(-t, v+r)\}c^2 e^{-c(u-v)}\, du\, dv$$

$$= \int_{-r-t}^{0} \int_{v+2r}^{\infty} (v+r)c^2 e^{-c(u-v)}\, du\, dv$$

$$+ \left(\int_{-2r}^{-t-r} \int_{v+2r}^{\infty} + \int_{-\infty}^{-2r} \int_{0}^{\infty} \right)(-t)c^2 e^{-c(u-v)}\, du\, dv$$

$$= \frac{c}{2}(r-t)^2 e^{-2cr} - te^{-2cr},$$

and for $t > r$

$$\iint\limits_{\substack{v<u-2r \\ v<0<u}} \{\max(-t, v+r)\}c^2 e^{-c(u-v)}\, du\, dv$$

$$= \left(\int_{-2r}^{0} \int_{v+2r}^{\infty} + \int_{-r-t}^{-2r} \int_{0}^{\infty} \right)(v+r)c^2 e^{-c(u-v)}\, du\, dv$$

$$+ \int_{-\infty}^{-r-t} \int_{0}^{\infty} (-t)c^2 e^{-c(u-v)}\, du\, dv$$

$$= -\left(r + \frac{1}{c}\right)e^{-2cr} + \frac{1}{c}e^{-c(r+t)}.$$

From (7.4.4) it follows that

$$p_4 = \begin{cases} \sqrt{\dfrac{2I}{\pi n}}\left\{t - \dfrac{c}{2}(t-r)^2\right\}e^{-2cr} & \text{for } t \leq r, \\[3mm] \sqrt{\dfrac{2I}{\pi n}}\left\{re^{-2cr} - \dfrac{1}{c}\left(e^{-c(t+r)} - e^{-2cr}\right)\right\} & \text{for } t > r. \end{cases}$$

From (7.4.2) and cases (i) to (iv) we have

$$(7.4.5) \quad P_0^n\left\{n|\hat{\theta}_{MP}^r| \leq t\right\}$$

$$= \begin{cases} 1 - e^{-2ct} + \dfrac{4ht}{\sqrt{2\pi In}}e^{-2cr} + \sqrt{\dfrac{2I}{\pi n}}\left\{t - \dfrac{c}{2}(t-r)^2\right\}e^{-2cr} + o\left(\dfrac{1}{\sqrt{n}}\right) \\ \hspace{8cm} \text{for } t \leq r, \\[3mm] 1 - e^{-c(r+t)} + \dfrac{2h}{\sqrt{2\pi In}}(r+t)e^{-c(r+t)} + \sqrt{\dfrac{2I}{\pi n}}te^{-c(r+t)} + o\left(\dfrac{1}{\sqrt{n}}\right) \\ \hspace{8cm} \text{for } t > r. \end{cases}$$

If $h = 0$ and $t = r$, then

$$P_0 \left\{ n \left| \hat{\theta}^r_{MP} \right| \le r \right\} = 1 - e^{-2cr} + \sqrt{\frac{2I}{\pi n}} r e^{-2cr} + o\left(\frac{1}{\sqrt{n}} \right),$$

hence it follows that the ACP of the MPE $\hat{\theta}^r_{MP}$ coincides with the bound given by Theorem 7.3.1 up to the order $o(n^{-1/2})$ at the point r. Thus we complete the proof.

7.5. EXAMPLES

The cases of the uniform and the truncated normal distributions are treated.

Example 7.5.1 (Uniform case). Let $f_0(x)$ be the following density :

$$f_0(x) = \begin{cases} 1/2 & \text{for} \quad |x| < 1; \\ 0 & \text{for} \quad |x| \ge 1. \end{cases}$$

Then we have

$$f_0(-1+0) = f_0(1-0) = c = \frac{1}{2}; \quad -f_0'(-1+0) = f_0'(1-0) = h = 0; \quad I = 0.$$

From Theorem 7.3.1 it follows that for any $\hat{\theta}_n \in A_2$, any $\theta \in \Theta$ and any $t > 0$

$$P^n_\theta \left\{ n \left| \hat{\theta}_n - \theta \right| \le t \right\} \le 1 - e^{-t} + \frac{t^2}{2n} e^{-t} + o\left(\frac{1}{n} \right).$$

If $L(u) = u^2$, then it follows from Theorem 7.2.3 that

$$P^n_\theta \left\{ n \left| \hat{\theta}_{GB} - \theta \right| \le t \right\} = 1 - e^{-t} + \frac{t^2}{2n} e^{-t} + o\left(\frac{1}{n} \right)$$

for all $t > 0$ and all $\theta \in \Theta$, since $h^+ = h^- = 0$. Hence $\hat{\theta}_{GB}$ is second order two-sided asymptotically efficient.

Example 7.5.2 (Symmetrically truncated normal case). Let $f_0(x)$ be the following density:

$$f_0(x) = \begin{cases} k e^{-x^2/2} & \text{for} \quad |x| < 1; \\ 0 & \text{for} \quad |x| \ge 1, \end{cases}$$

where k is some positive constant. Then we have

$$f_0(-1+0) = f_0(1-0) = k e^{-1/2} = c \quad \text{(say)},$$
$$- f_0'(-1+0) = f_0'(1-0) = h = -c.$$

From Theorem 7.3.1 it follows that for any $\hat{\theta}_n \in A_2$, any $\theta \in \Theta$ and any $t > 0$

$$P^n_\theta \left\{ n \left| \hat{\theta}_n - \theta \right| \le t \right\} \le 1 - e^{-2ct} + \sqrt{\frac{2I}{\pi n}} t e^{-2ct} + \frac{2}{n} c(c+1) t^2 e^{-2ct} + o\left(\frac{1}{n} \right),$$

where $c \fallingdotseq 0.36$ and $I \fallingdotseq 0.28$. If $L(u) = u^2$, then it follows from Theorem 7.2.3 that for the generalized Bayes estimator $\hat{\theta}_{GB}$

$$P_\theta^n \left\{ n \left| \hat{\theta}_{GB} - \theta \right| \le t \right\} \fallingdotseq 1 - e^{-2ct} + \frac{1}{n} e^{-2ct} \left(0.59t + 0.69t^2 - 0.11t^3 - 0.01t^4 \right)$$
$$+ o\left(\frac{1}{n} \right)$$

for all $t > 0$ and all $\theta \in \Theta$, $h^- = f_0'(-1 + 0) - f_0'(1 - 0) = 2c \fallingdotseq 0.72$ and $h^+ = 0$. Hence $\hat{\theta}_{GB}$ is first order two-sided asymptotically efficient but not 3/2-th order two-sided asymptotically efficient. From Remark 7.4.1 it is also seen that the MPE $\hat{\theta}_{MP}^r$ is not first order two-sided asymptotically efficient and does not attain the bound up to the order $o(n^{-1/2})$ at the point r since $h = -c < 0$ (see also Akahira and Takeuchi (1981, Section 3.3)).

7.6. SOME REMARKS

In Section 7.4, following the paper of Akahira (1991a), we consider the estimation problem on independent and identically distributed observations from a location parameter family generated by a density which is positive and symmetric on a finite interval, with a jump and a nonnegative right differential coefficient at the left endpoint. It is shown in Theorem 7.4.1 that the MPE is 3/2-th order two-sided asymptotically efficient at any fixed point in the sense that it has the most concentration probability around the true parameter at the point in the class of 3/2-th order AMU estimators only when the right differential coefficient vanishes at the left endpoint. The second order upper bound for the concentration probability of second order AMU estimators is also given in the paper. Further, it is shown in the paper that the MPE is second order two-sided asymptotically efficient at any fixed point in the above case only.

In the location parameter case the MPE is constructed from the extreme statistics and the maximum likelihood estimator, which is closely connected with the higher order sufficiency. Indeed, it is shown by Akahira (1991b) that, in one-directional family of distribution including the above location parameter family, the second order asymptotically sufficient statistics are given by the extreme statistics and an asymptotically ancillary statistic. The second order asymptotic sufficiency is also investigated from the viewpoint of the asymptotic loss of information using the amount (3.3.1) of information in Akahira (1994b). Further related results can be found in Akahira (1993), Blyth (1982) and Ibragimov-Has'minskii (1981).

SUPPLEMENT

THE BOUND FOR THE ASYMPTOTIC DISTRIBUTION OF ESTIMATORS WHEN THE MAXIMUM ORDER OF CONSISTENCY DEPENDS ON THE PARAMETER

S.1. ORDER OF CONSISTENCY DEPENDING ON THE PARAMETER

In the regular case it is known that the order of consistency is equal to $\sqrt{n}$, but in the non-regular case it is not always so, e.g. $n^{1/\alpha}$ $(0 < \alpha < 2)$, $\sqrt{n \log n}$ etc., which are independent of the unknown parameter (see Section 3.5 and also Akahira, 1975a; Akahira and Takeuchi, 1981; Vostrikova, 1984). In both cases the order of consistency is independent of θ. However, the order of consistency may depend on it in the case of the unstable and explosive process. Here a discussion on that will be done.

In the autoregressive process $\{X_t\}$ with $X_t = \theta X_{t-1} + U_t$ $(t = 1, 2, \ldots, T, \ldots)$, where $\{U_t\}$ is a sequence of independently and identically distributed random variables and $X_0 = 0$, it is known that the asymptotic distribution of the least squares estimator of θ is normal with the order $\sqrt{T}$ of consistency for $|\theta| < 1$, Cauchy with its order $|\theta|^T$ for $|\theta| > 1$ (e.g. White, 1958; Anderson, 1959) and some one with its order T for $|\theta| = 1$ (Rao, 1978). Further, in the case when $|\theta| < 1$ the asymptotic efficiency of estimators was studied by Akahira (1976a, 1982b), and, for the ARMA process, by Kabaila (1983), Taniguchi (1991) and others. (The limiting distributions of the least squares estimators with some orders of convergence are discussed by Yajima (1992) for misspecified models.)

Let $(\mathcal{X}, \mathcal{B})$ be a sample space and Θ be a parameter space, which is assumed to be an open set in Euclidean 1-space $\mathbf{R}^1$. We shall denote by $(\mathcal{X}^{(T)}, \mathcal{B}^{(T)})$ the T-fold direct product of $(\mathcal{X}, \mathcal{B})$. For each $T = 1, 2, \ldots$, the points of $\mathcal{X}^{(T)}$ will be denoted by $\mathbf{x}_T = (x_1, \ldots, x_T)$. We consider a sequence of classes of probability measures $\{P_{\theta,T} \mid \theta \in \Theta\}$ $(T = 1, 2, \ldots)$ each defined on $(\mathcal{X}^{(T)}, \mathcal{B}^{(T)})$ such that for each $T = 1, 2, \ldots$ and each $\theta \in \Theta$ the following holds :

$$P_{\theta,T}\left(B^{(T)}\right) = P_{\theta,T+1}\left(B^{(T)} \times \mathcal{X}\right)$$

for all $B^{(T)} \in \mathcal{B}^{(T)}$.

An estimator of θ is defined to be a sequence $\{\hat{\theta}_T\}$ of $\mathcal{B}^{(T)}$-measurable functions. For simplicity we may denote an estimator $\hat{\theta}_T$ instead of $\{\hat{\theta}_T\}$. For an increasing sequence $\{c_T\}$ (c_T tending to infinity) an estimator $\hat{\theta}_T$ is called consistent with order $\{c_T\}$ (or $\{c_T\}$-consistent for short) if for every $\vartheta \in \Theta$, there exists a sufficiently small positive number δ such that

$$\lim_{L\to\infty} \varlimsup_{T\to\infty} \sup_{\theta\,:\,|\theta-\vartheta|<\delta} P_{\theta,T}\left\{ c_T \left|\hat{\theta}_T - \theta\right| \geq L \right\} = 0.$$

If c_T can not be decided independently of θ, we may change $c_T(\theta)$ instead of c_T in the above definition. However, in such a definition we shall not be able to determine uniquely the value of $c_T(\theta)$ at a specified point θ_0. Then a similar phenomenon to "superefficiency" happens. Indeed, if $\hat{\theta}_T$ has an order c_T of consistency independent of θ, then for a specified value θ_0 we define an estimator $\hat{\theta}_T^*$ as

$$\hat{\theta}_T^* = \begin{cases} d_T^{-1}\left(\hat{\theta}_T - \theta_0\right) + \theta_0 & \text{for} \quad \left|\hat{\theta}_T - \theta_0\right| \leq c_T^{-1/2}; \\ \hat{\theta}_T & \text{for} \quad \left|\hat{\theta}_T - \theta_0\right| > c_T^{-1/2}, \end{cases}$$

where, for each $T = 1, 2, \ldots$, $c_T > 1$ and d_T is a constant with $d_T > 1$. We define $c_T^*(\theta)$ as follows :

$$c_T^*(\theta_0) = c_T d_T$$

and for any $\theta \neq \theta_0$

$$c_T^*(\theta) = \begin{cases} c_T^{1/2} & \text{for} \quad |\theta - \theta_0| \leq c_T^{-1/2}; \\ c_T\left\{ 1 - \dfrac{c_T^{-1/2}}{|\theta - \theta_0|} \right\} & \text{for} \quad |\theta - \theta_0| > c_T^{-1/2}. \end{cases}$$

Case (i) $\theta = \theta_0$. Since $\left|\hat{\theta}_T - \theta_0\right| \leq c_T^{-1/2}$ implies $c_T^*(\theta_0)\left|\hat{\theta}_T^* - \theta_0\right| = c_T d_T \left|\hat{\theta}_T^* - \theta_0\right| = c_T \left|\hat{\theta}_T - \theta_0\right|$, it follows that

$$P_{\theta_0,T}\left\{ c_T^*(\theta_0)\left|\hat{\theta}_T^* - \theta_0\right| \geq L \right\} \leq P_{\theta_0,T}\left\{ c_T\left|\hat{\theta}_T - \theta_0\right| \geq L \right\} + P_{\theta_0,T}\left\{ \left|\hat{\theta}_T - \theta_0\right| > c_T^{-1/2} \right\}$$

$$\leq P_{\theta_0,T}\left\{ c_T\left|\hat{\theta}_T - \theta_0\right| \geq \frac{L}{2} \right\}$$

$$+ P_{\theta_0,T}\left\{ c_T\left|\hat{\theta}_T - \theta_0\right| \geq c_T^{1/2} \right\}.$$

Case (ii) $\theta \neq \theta_0$. In the case when $|\theta - \theta_0| \leq c_T^{-1/2}$, $\left|\hat{\theta}_T - \theta_0\right| \leq c_T^{-1/2}$ implies $c_T^*(\theta)\left|\hat{\theta}_T^* - \theta\right| \leq c_T^{1/2}\left|\hat{\theta}_T^* - \theta\right| \leq c_T^{1/2} d_T^{-1}\left|\hat{\theta}_T - \theta_0\right| + c_T^{1/2}|\theta - \theta_0| \leq d_T^{-1} + 1 < 2$, and $\left|\hat{\theta}_T - \theta_0\right| > c_T^{-1/2}$ implies $c_T^*(\theta)\left|\hat{\theta}_T^* - \theta\right| = c_T^{1/2}\left|\hat{\theta}_T^* - \theta\right| \leq c_T\left|\hat{\theta}_T - \theta\right|$. Then we have for $L > 2$

$$P_{\theta,T}\left\{ c_T^*(\theta)\left|\hat{\theta}_T^* - \theta\right| \geq L \right\} \leq P_{\theta,T}\left\{ c_T\left|\hat{\theta}_T - \theta\right| \geq L \right\}.$$

In the case when $|\theta - \theta_0| > c_T^{-1/2}$, $\left|\hat{\theta}_T - \theta_0\right| \leq c_T^{-1/2}$ implies

$$
\begin{aligned}
c_T^*(\theta)\left|\hat{\theta}_T^* - \theta\right| &= c_T^*(\theta)\left|d_T^{-1}\left(\hat{\theta}_T - \theta_0\right) + \theta_0 - \theta\right| \\
&\leq c_T^*(\theta)d_T^{-1}\left|\hat{\theta}_T - \theta_0\right| + c_T^*(\theta)|\theta - \theta_0| \\
&\leq c_T^*(\theta)d_T^{-1}c_T^{-1/2} + c_T^*(\theta)|\theta - \theta_0| \\
&\leq c_T^*(\theta)\left(1 + d_T^{-1}\right)|\theta - \theta_0| \\
&\leq \left(1 + d_T^{-1}\right)c_T\left(|\theta - \theta_0| - c_T^{-1/2}\right) \\
&\leq \left(1 + d_T^{-1}\right)c_T\left|\hat{\theta}_T - \theta\right| \\
&< 2c_T\left|\hat{\theta}_T - \theta\right|,
\end{aligned}
$$

since $\left|\hat{\theta}_T - \theta\right| \geq |\theta - \theta_0| - c_T^{-1/2}$, and $\left|\hat{\theta}_T - \theta_0\right| > c_T^{-1/2}$ implies

$$
c_T^*(\theta)\left|\hat{\theta}_T^* - \theta\right| = c_T\left\{1 - \frac{c_T^{-1/2}}{|\theta - \theta_0|}\right\}\left|\hat{\theta}_T - \theta\right| \leq c_T\left|\hat{\theta}_T - \theta\right|.
$$

Then we have for $L > 2$

$$
P_{\theta,T}\left\{c_T^*(\theta)\left|\hat{\theta}_T^* - \theta\right| \geq L\right\} \leq 2P_{\theta,T}\left\{c_T\left|\hat{\theta}_T - \theta\right| \geq \frac{L}{2}\right\}.
$$

Hence in both cases (i) and (ii) we obtain for $L > 2$

$$
P_{\theta,T}\left\{c_T^*(\theta)\left|\hat{\theta}_T^* - \theta\right| \geq L\right\}
$$
$$
\leq 2P_{\theta,T}\left\{c_T\left|\hat{\theta}_T - \theta\right| \geq \frac{L}{2}\right\} + P_{\theta_0,T}\left\{c_T\left|\hat{\theta}_T - \theta_0\right| \geq c_T^{1/2}\right\}.
$$

Letting $T \to \infty$ and $L \to \infty$, we see that the right-hand side of the above inequality tends to zero locally uniformly. Since $\{d_T\}$ can be a sequence tending to infinity as $T \to \infty$ and for any $\theta \neq \theta_0$, $c_T^*(\theta)/c_T \to 1$ as $T \to \infty$, it is possible to make the order of convergence arbitrarily large at $\theta = \theta_0$. Hence we can not decide uniquely the value $c_T(\theta)$ at the point.

A $\{c_T\}$-consistent estimator $\hat{\theta}_T$ is defined to be asymptotically median unbiased (AMU) if for any $\vartheta \in \Theta$ there exists a positive number δ such that

$$
\lim_{T \to \infty} \sup_{\theta:|\theta - \vartheta| < \delta}\left|P_{\theta,T}\left\{\hat{\theta}_T \leq \theta\right\} - \frac{1}{2}\right| = 0,
$$
$$
\lim_{T \to \infty} \sup_{\theta:|\theta - \vartheta| < \delta}\left|P_{\theta,T}\left\{\hat{\theta}_T \geq \theta\right\} - \frac{1}{2}\right| = 0.
$$

Let $\hat{\theta}_T$ be an AMU estimator satisfying the following :

(S.1.1)
$$\varliminf_{T\to\infty}\left[P_{\theta,T}\left\{c_T\left(\hat{\theta}_T-\theta\right)\leq t\right\}-\frac{1}{2}\right]>0 \qquad \text{for some} \quad t>0,$$
$$\varlimsup_{T\to\infty}\left[P_{\theta,T}\left\{c_T\left(\hat{\theta}_T-\theta\right)\leq t\right\}-\frac{1}{2}\right]<0 \qquad \text{for some} \quad t<0.$$

If for any sequence $\{c'_T\}$ of order of consistency with (S.1.1), $\varlimsup_{T\to\infty} c'_T/c_T < \infty$, then $\{c_T\}$ is called the maximum order of consistency. In most cases the maximum order is uniquely determined, but there may not exist an estimator which satisfies (S.1.1) and uniformly attain the bound of

$$P_{\theta,T}\left\{c_T\left(\hat{\theta}_T-\theta\right)\leq t\right\}$$

in the class of AMU estimators with (S.1.1).

S.2. THE BOUND FOR THE ASYMPTOTIC DISTRIBUTION OF AMU ESTIMATORS IN THE AUTOREGRESSIVE PROCESS

We consider a simple autoregressive (AR) process $\{X_t\}$ which is defined by $X_t = \theta X_{t-1} + U_t$ $(t = 1, 2, \ldots)$, where $\{U_t\}$ is a sequence of independently, identically and normally distributed random variables with mean 0 and variance 1 and $X_0 = 0$.

In the subsequent discussion we shall treat the unstable case $|\theta| = 1$ and the explosive case $|\theta| > 1$ and obtain the asymptotic means and variances of a log-likelihood ratio test statistic under the null and the alternative hypothesis. We also discuss its asymptotic distribution. Then it is known that the order $\{c_T(\theta)\}$ of consistency is given by

$$c_T(\theta) = \begin{cases} |\theta|^T & \text{for} \quad |\theta| > 1; \\ T & \text{for} \quad |\theta| = 1, \end{cases}$$

(see, e.g. Anderson, 1959; Rao, 1978). Letting $|\theta_0| \geq 1$, we deal with the problem of testing the hypothesis $H : \theta = \theta_0 + (u/c_T(\theta_0))$ against the alternative $K : \theta = \theta_0$. Putting $\theta_1 = \theta_0 + (u/c_T(\theta_0))$ we consider the log-likelihood ratio statistic L_T given by

(S.2.1)
$$L_T = (\theta_0 - \theta_1)\left(\sum_{t=2}^{T} X_t X_{t-1} - \frac{\theta_0 + \theta_1}{2}\sum_{t=2}^{T} X_{t-1}^2\right).$$

Then we have the following.

Theorem S.2.1. *In the AR process the asymptotic means and variances of the log-likelihood ratio statistic L_T under $H : \theta = \theta_1$ and $K : \theta = \theta_0$ for $|\theta_0| \geq 1$ are given by*

the following table.

	$\lvert\theta_0\rvert > 1$	$\lvert\theta_0\rvert = 1$
$E_{\theta_0}(L_T)$	$\dfrac{u^2}{2\left(\theta_0^2 - 1\right)^2} + o(1)$	$\dfrac{u^2}{4} + o(1)$
$E_{\theta_1}(L_T)$	$-\dfrac{u^2}{2\left(\theta_0^2 - 1\right)^2} + o(1)$	$-\dfrac{1}{8}\left(e^{2u} - 1\right) + \dfrac{u}{4} + o(1)$
$V_{\theta_0}(L_T)$	$\dfrac{u^2}{\left(\theta_0^2 - 1\right)^2} + \dfrac{u^4}{2(\theta_0^2 - 1)^4} + o(1)$	$\dfrac{u^2}{2} - \dfrac{u^3}{3} + \dfrac{u^4}{12} + o(1)$
$V_{\theta_1}(L_T)$	$\dfrac{u^2}{\left(\theta_0^2 - 1\right)^2} + \dfrac{u^4}{2(\theta_0^2 - 1)^4} + o(1)$	$\dfrac{1}{4}\left(e^{2u} - 1\right) - \dfrac{u}{2} + O(1)$

Table S.2.1.

Proof of Theorem S.2.1. For each $t = 0, 1, 2, \ldots$, we put $\sigma_t^2 = V_\theta(X_t)$. Then, for each $k = 1, \ldots, t$, we have

$$(S.2.2) \qquad E_\theta\left(X_t X_{t-k}\right) = \theta E_\theta\left(X_{t-1} X_{t-k}\right) = \cdots = \theta^k V_\theta\left(X_{t-k}\right) = \theta^k \sigma_{t-k}^2 \quad \text{(say)}.$$

Since for each $t = 1, 2, \ldots$,

$$V_\theta\left(X_t\right) = \theta^2 V_\theta\left(X_{t-1}\right) + V\left(U_t\right),$$

it follows that

$$(S.2.3) \qquad\qquad \sigma_t^2 = \theta^2 \sigma_{t-1}^2 + 1, \qquad (t = 1, 2, \ldots),$$

where $\sigma_0^2 = 0$. First we obtain

$$(S.2.4) \quad V_\theta(L_T) = (\theta_0 - \theta_1)^2 \left\{ V_\theta\left(\sum_{t=2}^{T} U_t X_{t-1}\right) \right.$$

$$+ (2\theta - \theta_0 - \theta_1)\, Cov_\theta\left(\sum_{t=2}^{T} U_t X_{t-1}, \sum_{t=2}^{T} X_{t-1}^2\right)$$

$$\left. + \left(\theta - \frac{\theta_0 + \theta_1}{2}\right)^2 V_\theta\left(\sum_{t=2}^{T} X_{t-1}^2\right) \right\}.$$

Then we have

$$(S.2.5) \quad V_\theta\left(\sum_{t=2}^{T} U_t X_{t-1}\right) = \sum_{t=2}^{T} V_\theta\left(U_t X_{t-1}\right) + 2\sum_{t<t'}\sum Cov_\theta\left(U_t X_{t-1}, U_{t'} X_{t'-1}\right)$$

$$= \sum_{t=2}^{T} \sigma_{t-1}^2.$$

We also obtain by (S.2.2) and (S.2.3)

(S.2.6) $$V_\theta \left(\sum_{t=2}^{T} X_{t-1}^2 \right) = 2 \sum_{t=2}^{T} \sigma_{t-1}^4 + 4 \sum_{t<t'} \sum \sigma_{t-1}^4 \theta^{2(t'-t)},$$

(S.2.7) $$Cov_\theta \left(\sum_{t=2}^{T} U_t X_{t-1}, \sum_{t=2}^{T} X_{t-1}^2 \right) = 2 \sum_{t<t'} \sum \sigma_{t-1}^2 \theta^{t'-t}.$$

Since by (S.2.2)

$$E_\theta \left(\sum_{t=2}^{T} U_t X_{t-1} \right) = 0, \quad E_\theta \left(\sum_{t=2}^{T} X_{t-1}^2 \right) = \sum_{t=1}^{T-1} \sigma_t^2,$$

it follows from (S.2.1) that

(S.2.8) $$E_\theta(L_T) = (\theta_0 - \theta_1) \left(\theta - \frac{\theta_0 + \theta_1}{2} \right) \sum_{t=1}^{T-1} \sigma_t^2.$$

Since by (S.2.3)

$$\sigma_t^2 = 1 + \theta^2 + \cdots + \theta^{2(t-1)} = \begin{cases} \dfrac{\theta^{2t} - 1}{\theta^2 - 1} & \text{for } |\theta| \neq 1, \\ t & \text{for } |\theta| = 1, \end{cases}$$

it follows that

(S.2.9) $$\sum_{t=1}^{T-1} \sigma_t^2 = \begin{cases} \dfrac{\theta^{2T} - 1}{(\theta^2 - 1)^2} - \dfrac{T}{\theta^2 - 1} & \text{for } |\theta| \neq 1, \\ \dfrac{T(T-1)}{2} & \text{for } |\theta| = 1. \end{cases}$$

Case (I): $\theta_0 > 1$. By (S.2.8) and (S.2.9) we have

(S.2.10) $$E_{\theta_0}(L_T) = \frac{u^2}{2 (\theta_0^2 - 1)^2} + o(1),$$

(S.2.11) $$E_{\theta_1}(L_T) = -\frac{u^2}{2 (\theta_0^2 - 1)^2} + o(1),$$

since $\theta_0 - \theta_1 = -u\theta_0^{-T}$ and $\theta_0 + \theta_1 = 2\theta_0 + u\theta_0^{-T}$. It also follows from (S.2.3) to (S.2.7) and (S.2.9) that

(S.2.12) $$V_{\theta_0}(L_T) = \frac{u^2}{(\theta_0^2 - 1)^2} + \frac{u^4}{2(\theta_0 - 1)^4} + o(1).$$

Similarly we have

$$(S.2.13) \qquad V_{\theta_1}(L_T) = \frac{u^2}{(\theta_0^2 - 1)^2} + \frac{u^4}{2(\theta_0 - 1)^4} + o(1).$$

In the case when $\theta_0 < -1$, we have similar asymptotic means and variances of L_T under H and K to those of this case.

Case (II) : $\theta_0 = 1$. Since $\theta_1 = 1 + (u/T)$, it follows from (S.2.8) and (5.2.9) that

$$(S.2.14) \qquad E_1(L_T) = \frac{u^2}{4} + o(1).$$

From (S.2.8) and (S.2.9), we obtain under $H : \theta = \theta_1 = 1 + (u/T)$

$$(S.2.15) \qquad E_{\theta_1}(L_T) = -\frac{u^2}{2T^2} \left\{ \frac{\theta_1^{2T} - 1}{(\theta_1^2 - 1)^2} - \frac{T}{\theta_1^2 - 1} \right\}.$$

Since

$$\frac{1}{\theta_1^2 - 1} = \frac{T}{2u + (u^2/T)},$$

it follows from (S.2.15) that

$$(S.2.16) \qquad E_{\theta_1}(L_T) = -\frac{u^2}{2T^2} \left[\left\{ \left(1 + \frac{u}{T}\right)^{2T} - 1 \right\} \frac{T^2}{\left(2u + \dfrac{u^2}{T}\right)^2} - \frac{T^2}{2u + \dfrac{u^2}{T}} \right]$$

$$= -\frac{1}{8} \left(e^{2u} - 1\right) + \frac{u}{4} + o(1).$$

We also have the asymptotic variances as

$$(S.2.17) \quad V_1(L_T) = \frac{u^2}{2} - \frac{u^3}{3} + \frac{u^4}{12} + o(1),$$

$$(S.2.18) \quad V_{\theta_1}(L_T) = \frac{1}{4} \left(e^{2u} - 1\right) - \frac{u}{2} + \frac{1}{32} \left\{ e^{4u} + 4(2u - 3)e^{2u} + 12u + 15 \right\} + o(1)$$

$$= \frac{1}{4} \left(e^{2u} - 1\right) - \frac{u}{2} + O(1).$$

For the case when $\theta_0 = -1$, in a similar way to the case $|\theta_0| > 1$, we can obtain the asymptotic means and variances of L_T under H and K. Thus we complete the proof.

Next we needs to consider the asymptotic distribution of L_T under H and K. Using the discussion by Basawa and Brockwell (1984, page 165), we see that L_T converges in distribution to

$$\{a_{\theta_i}(u)Y + b_{\theta_i}(u)Z\} Y \qquad (i = 0, 1)$$

as $T \to \infty$ under H and K, respectively, where for each $i = 0, 1$, $a_{\theta_i}(u)$ and $b_{\theta_i}(u)$ are constants, and where Y and Z are mutually, independently and normally distributed random variables with mean 0 and variance 1. Note that the asymptotic distribution of L_T is not the normal as in the regular cases, which means that the bound for the asymptotic distribution of AMU estimators can not be expressed in terms of the normal distribution, but from the above formula we may obtain the bound for the asymptotic distribution of the all AMU estimators. It has been shown that the maximum likelihood estimator and other "usual" asymptotically efficient estimators do not uniformly attain the bound.

REFERENCES

Akahira, M. (1975a). Asymptotic theory for estimation of location in non-regular cases, I: Order of convergence of consistent estimators. *Rep. Stat. Appl. Res., JUSE*, **22**, 8–26.

Akahira, M. (1975b). Asymptotic theory for estimation of location in non-regular cases, II: Bounds of asymptotic distributions of consistent estimators. *Rep. Stat. Appl. Res., JUSE*, **22**, 99–115.

Akahira, M. (1976a). On the asymptotic efficiency of estimators in an autoregressive process. *Ann. Inst. Statist. Math.*, **28**, 35–48.

Akahira, M. (1976b). A remark on asymptotic sufficiency of statistics in non-regular cases. *Rep. Univ. Electro-Comm.*, **27**, 125–128.

Akahira, M. (1982a). Asymptotic optimality of estimators in non-regular cases. *Ann. Inst. Statist. Math.*, **34**, part A, 69–82.

Akahira, M. (1982b). Second order asymptotic optimality of estimators in an autoregressive process with unknown mean. *Selecta Statistica Canadiana* **6**, 19–36.

Akahira, M. (1986). *The Structure of Asymptotic Deficiency of Estimators*. Queen's Papers in Pure and Appl. Math., 75, Queen's Univ. Press, Kingston, Ontario, Canada.

Akahira, M. (1987). Second order asymptotic comparison of estimators of a common parameter in the double exponential case. *Ann. Inst. Statist. Math.*, **39**, 25–36.

Akahira, M. (1988a). Second order asymptotic optimality of estimators for a density with finite cusps. *Ann. Inst. Statist. Math.*, **40**, 311–328.

Akahira, M. (1988b). Second order asymptotic properties of the generalized Bayes estimators for a family of non-regular distributions. In : *Statistical Theory and Data Analysis II*, (K. Matusita, Ed.), North-Holland, Amsterdam, 87–100.

Akahira, M. (1990). Second order asymptotic comparison of the discretized likelihood estimator with asymptotically efficient estimators in the double exponential case. *Metron*, **48**, 5–17.

Akahira, M. (1991a). The 3/2th and 2nd order asymptotic efficiency of maximum probability estimators in non-regular cases. *Ann. Inst. Statist. Math.*, **43**, 181–195.

Akahira, M. (1991b). Second order asymptotic sufficiency for a family of distributions with one-directionality. *Metron*, **49**, 133–143.

Akahira, M. (1993). Asymptotics on the statistics for a family of non-regular distributions. *Stat. Sci. & Data Anal.*, (K. Matusita et al. Eds.), VSP Internat. Sci. Publ., Zeist (Netherlands), 357–364.

Akahira, M. (1994a). The amount of information and the bound for the order of consistency for a location parameter family of densities. *To appear in the Proceedings of the 2nd Gauss Symposium*, de Gruyter.

Akahira, M. (1994b). Loss of information of the statistics for a family of non-regular distributions. *In revision in the Ann. Inst. Statist. Math.*

Akahira, M., Puri, M. L. and Takeuchi, K. (1986). Bhattacharyya bound of variances of unbiased estimators in non-regular cases. *Ann. Inst. Statist. Math.*, **38**, 35–44.

Akahira, M. and Takeuchi, K. (1979). Discretized likelihood methods— Asymptotic properties of discretized likelihood estimators (DLE's). *Ann. Inst. Statist. Math.*, **31**, 39–56.

Akahira, M. and Takeuchi, K. (1981). *Asymptotic Efficiency of Statistical Estimators: Concepts and Higher Order Asymptotic Efficiency.* Lecture Notes in Statistics 7, Springer, New York.

Akahira, M. and Takeuchi, K. (1982). On asymptotic deficiency of estimators in pooled samples in the presence of nuisance parameters. *Statistics and Decisions* **1**, 17–38.

Akahira, M. and Takeuchi, K. (1985a). A note on the minimum variance unbiased estimation when the Fisher information is infinity. *Rep. Stat. Appl. Res., JUSE*, **32**, 17–22.

Akahira, M. and Takeuchi, K. (1985b). Estimation of a common parameter for pooled samples from the uniform distributions. *Ann. Inst. Statist. Math.*, **37**, Part A, 131–140.

Akahira, M. and Takeuchi, K. (1986). On the bound of the asymptotic distribution of estimators when the maximum order of consistency depends on the parameter. *Pub. Inst. Stat. Univ. Paris*, **31**, 1–16.

Akahira, M. and Takeuchi, K. (1987a). The lower bound for the variance of unbiased estimators for one-directional family of distributions. *Ann. Inst. Statist. Math.*, **39**, 593–610.

Akahira, M. and Takeuchi, K. (1987b). Locally minimum variance unbiased estimator in a discontinuous density function. *Metrika*, **34**, 1–15.

Akahira, M. and Takeuchi, K. (1989). Higher order asymptotics in estimation for two-sided Weibull type distributions. *Ann. Inst. Statist. Math.*, **41**, 725–752.

Akahira, M. and Takeuchi, K. (1990). Loss of information associated with the order statistics and related estimators in the double exponential distribution case. *Austral. J. Statist.*, **32**, 281–291.

Akahira, M. and Takeuchi, K. (1991a). A definition of information amount applicable to non-regular cases. *Journal of Computing and Information* **2**, 71–92.

Akahira, M. and Takeuchi, K. (1991b). Asymptotic efficiency of estimators for a location parameter family of densities with the bounded support. *Rep. Stat. Appl. Res., JUSE*, **38**, 1–9.

Akahira, M. and Takeuchi, K. (1993). Second order asymptotic bound for the variance of estimators for the double exponential distribution. *Stat. Sci. & Data Anal.*, (K. Matsusita et al. Eds.), VSP Internat. Sci. Publ., Zeist (Netherlands), 375–382.

Amari, S. (1985). *Differential-Geometrical Methods in Statistics*. Lecture Notes in Statistics 28, Springer, Berlin.

Anderson, T. W. (1959). On a asymptotic distributions of estimates of stochastic difference equations. *Ann. Math. Statist.*, **30**, 676–687.

Antoch, J. (1984). Behaviour of estimators of location in non-regular cases : A Monte Carlo study. In : *Asymptotic Statistics 2, Proceedings of the Third Prague Symposium on Asymptotic Statist.*, North-Holland, Amsterdam, 185–195.

Bahadur, R. R. (1960). On the asymptotic efficiency of tests and estimators. *Sankhyā*, **22**, 229–252.

Barankin, E. W. (1949). Locally best unbiased estimates. *Ann. Math. Statist.*, **20**, 477–501.

Barndorff-Nielsen, O. E. and Cox, D. R. (1994). *Inference and Asymptotics*. Chapman and Hall, London.

Basawa, I. V. and Brockwell, P. J. (1984). Asymptotic conditional inference for regular nonergodic models with an application to autoregressive processes. *Ann. Statist.*, **12**, 161–171.

Basu, D. (1955). On statistics independent of a complete sufficient statistic. *Sankhyā*, **15**, 377–380.

Bhattacharya, C. G. (1981). Estimation of a common location. *Commun. Statist.*, **A10** (10), 955–961.

Blyth, C. R. (1982). Maximum probability estimation in small samples. In : *A Festschrift for E. L. Lehmann* (P. J. Bickel et al. Eds), Wadsworth, Belmont, 83–96.

Boente, G. and Fraiman, R. (1988). On the asymptotic behaviour of general maximum likelihood estimates for the non-regular case under nonstandard conditions. *Biometrika* **75**, 45–56.

Cassel, C. M., Särndal, C. E., and Wretman, J. H. (1977). *Foundations of Inference in Servey Sampling*. Wiley, New York.

Chapman, D. G. and Robbins, H. (1951). Minimum variance estimation without regularity assumptions. *Ann. Math. Statist.*, **22**, 581–586.

Chatterji, S. D. (1982). A remark on the Cramér-Rao inequality. In : *Statistics and Probability : Essays in Honor of C. R. Rao*, North-Holland, New York, 193–196.

178

Chernoff, H. and Rubin, H. (1956). The estimation of the location of a discontinuity in density. *Proc. Third Berkeley Symp. Math. Statist. Prob.*, **1**, 17–37.

Cohen, A. (1976). Combining estimates of location. *J. Amer. Statist. Assoc.*, **71**, 172–175.

Daniels, H. E. (1961). The asymptotic efficiency of a maximum likelihood estimator. *Proc. Fourth Berkeley Symp. Math. Statist. Prob.*, **1**, 151–163.

Fend, A. V. (1959). On the attainment of the Cramér-Rao and Bhattacharyya bounds for the variance of an estimate. *Ann. Math. Statist.*, **30**, 381–388.

Fisher, R. A. (1922). On the mathematical foundations of theoretical statistics. *Philos. Trans. Roy. Soc. London, Ser. A*, **222**, 309–368.

Fisher, R. A. (1925). Theory of statistical estimation. *Proc. Cambridge Philos. Soc.*, **22**, 700–725.

Fisher, R. A. (1934). Two new properties of mathematical likelihood. *Proc. Roy. Soc. London*, A-**144**, 285–307.

Fisher, R. A. (1956). *Statistical Methods and Scientific Inference.* Oliver & Boyd, Edinburgh.

Fraser, D. A. S. and Guttman, I. (1952). Bhattacharyya bounds without regularity assumptions. *Ann. Math. Statist.*, **23**, 629–632.

Ghosh, J. K. (1994). *Higher Order Asymptotics.* NSF-CBMS Regional Conference Series Probab. and Statist., 4, Inst. of Math. Statist., Hayward, California.

Ghosh, J. K., Sinha, B. K. and Wieand, H. S. (1980). Second order efficiency of the mle with respect to any bounded bowl-shaped loss function. *Ann. Statist.*, **8**, 506–521.

Hall, P. (1982). On estimating the endpoint of a distribution. *Ann. Statist.*, **10**, 556–568.

Hammersley, J. M. (1950). On estimating restricted parameters. *J. Roy. Statist. Soc., Ser. B*, **12**, 192–240.

Hoeffding, W. and Wolfowitz, J. (1958). Distinguishability of sets of distributions. *Ann. Math. Statist.*, **29**, 700-718.

Ibragimov, I. A. and Has'minskii, R. Z. (1981). *Statistical Estimation : Asymptotic Theory.* Springer, New York.

Isii, K. (1964). Inequalities of the types of Chebyshev and Cramér-Rao and mathematical programming. *Ann. Inst. Statist. Math.*, **16**, 277–293.

Janssen, A. and Reiss, R.-D. (1988). Comparison of location models of Weibull type samples and extreme value process. *Probab. Th. Rel. Fields* **78**, 273–292.

Jurečková, J. (1981). Tail-behaviour of location estimators in non-regular cases. *Commentationes Mathematicae Universitatis Carolinae* **22**, 365–375.

Kabaila, P. (1983). On the asymptotic efficiency of the estimators of the parameters of an ARMA process. *Journal of Time Series Analysis*, 4, 37–47.

Khatri, C. G. (1980). Unified treatment of Cramér-Rao bound for the non-regular density functions. *J. Statist. Plann. Inference* 4, 75–79.

Kiefer, J. (1952). On minimum variance in non-regular estimation. *Ann. Math. Statist.*, 23, 627–629.

Kojima, Y., Morimoto, H. and Takeuchi, K. (1982). Two best unbiased estimators of normal integral mean. In : *Statistics and Probability : Essays in Honor of C. R. Rao*, North-Holland, New York, 429–441.

Kullback, S. (1959). *Information Theory and Statistics*. Wiley, New York.

LeCam, L. (1986). *Asymptotic Methods in Statistical Decision Theory*. Springer-Verlag, New York.

LeCam, L. (1990). On standard asymptotic confidence ellipsoids of Wald. *Internat. Statist. Rev.*, 58, 129–152.

Lehmann, E. L. (1983). *Theory of Point Estimation*. Wiley, New York.

Lindley, D. V. (1961). The use of prior probability distributions in statistical inference and decision. *Proc. Fourth Berkeley Symp. Math. Statist. Prob.*, 1, 453–468.

Lindley, D. V. (1972). *Bayesian Statistics, A Review*. SIAM, Philadelphia.

Matusita, K. (1955). Decision rules based on the distance for problems of fit, two-samples and estimation. *Ann. Math. Statist.*, 26, 631–640.

Mita, H. (1979). Asymptotic sufficiency of maximum likelihood estimator in a truncated location family. *Tokyo J. Math.*, 2, 323–335.

Móri, T. F. (1983). Note on the Cramér-Rao inequality in the non-regular case : The family of uniform distributions. *J. Statist. Plann. Inference*, 7, 353–358.

Morimoto, H. and Sibuya, M. (1967). Sufficient statistics and unbiased estimation of restricted selection parameter. *Sankhyā* (A) 27, 15–40.

Papaioannou, P. C. and Kempthorne, O. (1971). On statistical information theory and related measures of information. *Aerospace Research Laboratories*.

Pfanzagl, J. and Wefelmeyer, W. (1985). *Asymptotic Expansions for General Statistical Models*. Lecture Notes in Statistics 31, Springer, Berlin.

Polfeldt, T. (1970a). Asymptotic results in non-regular estimation. *Skand. Akt. Tidskr. Suppl.*, 1–2.

Polfeldt, T. (1970b). The order of the minimum variance in a non-regular case. *Ann. Math. Statist.*, 41, 667–672.

Prakasa Rao, B. L. S. (1968). Estimation of the location of the cusp of a continuous density. *Ann. Math. Statist.*, 39, 76–87.

Rao, C. R. (1952). Some theorems on minimum variance estimation. *Sankhyā* **12**, 27–42.

Rao, C. R. (1961). Asymptotic efficiency and limiting information. *Proc. Fourth Berkeley Symp. Math. Statist. Prob.*, **1**, 531–545.

Rao, C. R. (1973). *Linear Statistical Inference and Its Applications*. Wiley, New York.

Rao, M. M. (1978). Asymptotic distribution of an estimator of the boundary parameter of an unstable process. *Ann. Statist.*, **6**, 185–190. Correction, *ibid.*, **8**, 1404.

Sen, P. K. and Ghosh, B. K. (1976). Comparison of some bounds in estimation theory. *Ann. Statist.*, **4**, 755–765. Correction, *ibid.*, **5**, 593.

Smith, R. L. (1985). Maximum likelihood estimation in a class of non-regular cases. *Biometrika* **72**, 67–90.

Smith, R. L. (1989). A survey of nonregular problems. *Bull. Int. Statist. Inst.*, **53**, 353–372.

Stein, C. (1950). Unbiased estimates of minimum variance. *Ann. Math. Statist.*, **21**, 406–415.

Sugiura, N. and Naing, M. T. (1989). Improved estimators for the location of double exponential distribution. *Commun. Statist.* A– Theory Methods **18**, 541–554.

Takeuchi, K. (1961a). Some remarks on unbiased estimation in sampling from finite populations. *Rep. Stat. Appl. Res., JUSE*, **8**, 207–210.

Takeuchi, K. (1961b). On the fallacy of a theory of Gunner Blom. *Rep. Stat. Appl. Res., JUSE*, **9**, 34–35.

Takeuchi, K. and Akahira, M. (1979). Notes on non-regular asymptotic estimation —what "non-regularity" implies. *Rep. Univ. Electro-Comm.*, **30**, 63–66.

Takeuchi, K. and Akahira, M. (1986). A note on minimum variance. *Metrika*, **33**, 85–91.

Taniguchi, M. (1991). *Higher Order Asymptotic Theory for Time Series Analysis*. Lecture Notes in Statistics 67, Springer, New York.

Víncze, I. (1979). On the Cramér-Fréchet-Rao inequality in the non-regular case. In: *Contributions to Statistics. The Jaroslav Hájek Memorial Volume*, Academia, Prague, 253–262.

Vostrikova, L. Ju. (1984). On criteria for c_n-consistency of estimators. *Stochastics*, **11**, 265–290.

Weiss, L. (1979). Asymptotic sufficiency in a class of non-regular cases. *Selecta Statistica Canadiana* **5**, 141–150.

Weiss, L. and Wolfowitz, J. (1967). Maximum probability estimators. *Ann. Inst. Statist. Math.*, **19**, 193–206.

Weiss, L. and Wolfowitz, J. (1973). Maximum likelihood estimation of a translation parameter of a truncated distribution. *Ann. Statist.*, **1**, 944–947.

Weiss, L. and Wolfowitz, J. (1974). *Maximum Probability Estimators and Related Topics*. Lecture Notes in Math., 424, Springer, Berlin.

White, J. S. (1958). The limiting distribution of the serial correlation coefficient in the explosive case. *Ann. Math. Statist.*, **29**, 1188–1197.

Williamson, J. A. (1984). A note on the proof by H. E. Daniels of the asymptotic efficiency of a maximum likelihood estimator. *Biometrika* **71**, 651–653.

Woodroofe, M. (1972). Maximum likelihood estimation of a translation parameter of a truncated distribution. *Ann. Math. Statist.*, **43**, 113–122.

Woodroofe, M. (1974). Maximum likelihood estimation of translation parameter of truncated distribution II. *Ann. Statist.*, **2**, 474–488.

Yajima, Y. (1992). Asymptotic properties of estimators in incorrect ARMA models for long-memory time series. *New Directions in Time Series Analysis II* (D. Brillinger et al. Eds.), Springer, New York, 375–382.

Zacks, S. (1971). *The Theory of Statistical Inference*. Wiley, New York.

Subject Index

Lecture Notes in Statistics

For information about Volumes 1 to 20
please contact Springer-Verlag